Information and Instructions

This shop manual contains several sections each covering a specific group of wheel type tractors. The Tab Index on the preceding page can be used to locate the section pertaining to each group of tractors. Each section contains the necessary specifications and the brief but terse procedural data needed by a mechanic when repairing a tractor on which he has had no previous actual experience.

Within each section, the material is arranged in a systematic order beginning with an index which is followed immediately by a Table of Condensed Service Specifications. These specifications include dimensions, fits, clearances and timing instructions. Next in order of arrangement is the procedures paragraphs.

In the procedures paragraphs, the order of presentation starts with the front axle system and steering and proceeding toward the rear axle. The last paragraphs are devoted to the power take-off and power lift systems. Interspersed where needed are additional tabular specifications pertaining to wear limits, torquing, etc.

HOW TO USE THE INDEX

Suppose you want to know the procedure for R&R (remove and reinstall) of the engine camshaft. Your first step is to look in the index under the main heading of ENGINE until you find the entry "Camshaft." Now read to the right where under the column covering the tractor you are repairing, you will find a number which indicates the beginning paragraph pertaining to the camshaft. To locate this wanted paragraph in the manual, turn the pages until the running index appearing on the top outside corner of each page contains the number you are seeking. In this paragraph you will find the information concerning the removal of the camshaft.

More information available at haynes.com
Phone: 805-498-6703

T0256164

Haynes UK
Sparkford Nr Yeovil
Somerset BA22 7JJ England

Haynes North America, Inc
859 Lawrence Drive
Newbury Park
California 91320 USA

ISBN-10: 0-87288-420-1
ISBN-13: 978-0-87288-420-5

SHOP MANUAL
FORD
MODELS
1120—1220—1320—1520—1720—1920—2120

The tractor model number, serial number and engine number are stamped on an identification plate located on left side of transmission housing.

INDEX (By Starting Paragraph)

INDEX (CONT.)

INDEX (CONT.)

DUAL DIMENSIONS

This service manual provides specifications in both the Metric (SI) and U.S. Customary systems of measurement. The first specification is given in the measuring system used during manufacture, while the second specification (given in parenthesis) is the converted measurement. For instance, a specification of "0.28 mm (0.011 inch)" would indicate that the equipment was manufactured using the metric system of measurement and the U.S. Customary equivalent of 0.28 mm is 0.011 inch.

CONDENSED SERVICE DATA

	Models		
	1120	**1220**	**1320**
GENERAL			
Engine Make	———————— Shibaura ————————		
Engine Model	S723	S753	J823
Number of Cylinders	——————————3——————————		
Bore	72 mm	75 mm	82 mm
	(2.83 in.)	(2.95 in.)	(3.22 in.)
Stroke	72 mm	72 mm	80 mm
	(2.83 in.)	(2.83 in.)	(3.15 in.)
Displacement	879 cc	954 cc	1267 cc
	(53.6 cu. in.)	(58.2 cu. in.)	(77.3 cu. in.)
Compression Ratio	23:1	23:1	22:1
Electrical System			
Alternator	———————— 12 Volts, 35 Amps ————————		
Regulator	Mechanical	Mechanical	Solid State
Battery Ground Polarity	———————— Negative ————————		
TUNE-UP			
Firing Order	———————— 1-2-3 ————————		
Valve Clearance—Cold			
Intake	———————— 0.20 mm ————————		
	(0.008 in.)		
Exhaust	———————— 0.20 mm ————————		
	(0.008 in.)		
Valve Face Angle	———————— 45° ————————		
Valve Seat Angle	———————— 45° ————————		
Injection Timing—			
Static, BTDC	23°-24°	23°-24°	20°-21°
Timing Mark Location	————————Crankshaft Pulley————————		
Injector Opening Pressure	————————11760 kPa————————		
	(1705 psi)		
Governed Speeds—Engine Rpm			
Low Idle	———————— 800-900 ————————		
High Idle (No-Load)	————————2650-2700————————		
Rated (Full Load)	———————— 2500 ————————		
Rated Power at Pto	9.3 kW	10.8 kW	12.7 kW
	(12.5 hp)	(14.5 hp)	(17.0 hp)
SIZES—CLEARANCES			
Crankshaft Main Journal			
Diameter	———— 45.964-45.975 mm ————		57.957-57.97 mm
	(1.8096-1.810 in.)		(2.281-2.282 in.)

CONDENSED SERVICE DATA (CONT.)

	Models	
1120	**1220**	**1320**

SIZES—CLEARANCES (Cont.)

	1120	1220	1320
Main Bearing Radial Clearance		0.039-0.106 mm (0.0015-0.004 in.)	0.044-0.116 mm (0.0017-0.0045 in.)
Crankshaft End Play		0.05-0.30 mm (0.002-0.011 in.)	0.10-0.40 mm (0.004-0.016 in.)
Crankpin Diameter		38.964-38.975 mm (1.5340-1.5344 in.)	43.964-43.975 mm (1.730-1.731 in.)
Rod Bearing Radial Clearance		0.035-0.083 mm (0.001-0.003 in.)	0.035-0.083 mm (0.001-0.003 in.)
Connecting Rod Side Clearance		0.10-0.30 mm (0.004-0.012 in.)	0.10-0.30 mm (0.004-0.012 in)
Piston-to-Cylinder Bore Clearance	0.0575-0.0875 mm (0.0022-0.0034 in.)	0.0425-0.0765 mm (0.0016-0.0030 in.)	0.088-0.106 mm (0.0034-0.0041 in.)

SPECIAL TORQUES

	1120	1220	1320
Connecting Rod Caps		24-27 N·m (17-20 ft.-lbs.)	49-54 N·m (36-40 ft.-lbs.)
Crankshaft Main Bearing Holders		25-29 N·m (18-22 ft.-lbs.)	49-54 N·m (36-40 ft.-lbs.)
Main Bearing Holder Retaining Bolts		25-29 N·m (18-22 ft.-lbs.)	49-54 N·m (36-40 ft.-lbs.)
Crankshaft Pulley Retaining Nut		48-58 N·m (36-43 ft.-lbs.)	275-333 N·m (203-245 ft.-lbs.)
Flywheel Bolts		59-69 N·m (44-50 ft.-lbs.)	59-69 N·m (44-50 ft.-lbs.)
Cylinder Head Bolts		44-49 N·m (33-36 ft.-lbs.)	88-93 N·m (65-68 ft.-lbs.)

CAPACITIES

	1120	1220	1320
Cooling System		3.5 liters (3.7 U.S. qts.)	Note 1
Crankcase with Filter Change		3.3 liters (3.5 U.S. qts.)	4.5 liters (4.8 U.S. qts.)
Fuel Tank		18 liters (4.8 U.S. gals.)	27 liters (7.1 U.S. gals.)
Rear Axle and Transmission		16 liters (16.9 U.S. qts.)	22 liters (23.3 U.S. qts.)
Power Steering		1.8 liters (1.9 U.S. qts.)	
Front Wheel Drive Axle		1.8 liters (1.9 U.S. qts.)	2.8 liters (3 U.S. qts.)

Note 1: Model 1320 cooling system capacity is 4 liters (4.2 U.S. qts.) with gear transmission and 5 liters (5.3 U.S. qts.) with hydrostatic transmission.

CONDENSED SERVICE DATA

	Models			
	1520	**1720**	**1920**	**2120**
GENERAL				
Engine Make	————————— Shibaura —————————			
Engine Model	J843	N843	N844	T854 B
Number of Cylinders	————————3————————		——— 4 ———	
Bore	————— 84 mm ————— (3.307 in.)		84 mm (3.307 in.)	85 mm (3.346 in.)
Stroke	80 mm (3.150 in.)	90 mm (3.543 in.)	90 mm (3.543 in.)	100 mm (3.937 in.)
Displacement	1330 cc (81.1 cu. in.)	1496 cc (91.3 cu. in.)	1995 cc (121.7 cu. in.)	2268 cc (138.4 cu. in.)
Compression Ratio	22:1	22.5:1	19:1	18:1
Electrical System				
Alternator	————————— 12 Volt, 35 Amps —————————			
Regulator	———————Solid State, Integral———————			
Battery Ground Polarity	————————— Negative —————————			
TUNE-UP				
Firing Order	————— 1-2-3 —————		————— 1-3-4-2 —————	
Valve Clearance—Cold				
Intake	————— 0.2 mm ————— (0.008 in.)			0.3 mm (0.012 in.)
Exhaust	————— 0.2 mm ————— (0.008 in.)			0.3 mm (0.012 in.)
Valve Face Angle	————— 45° —————			
Valve Seat Angle	————— 45° —————			
Injection Timing-Static, BTDC	20°-21°	22°-23°	18°-19°	19.5°-20.5°
Timing Mark Location	————————— Crankshaft Pulley —————————			
Injector Opening Pressure	11760 kPa (1705 psi)	14825 kPa (2150 psi)	20590 kPa (2985 psi)	19615 kPa (2845 psi)
Governed Speeds-Engine Rpm				
Low Idle	—————800-900—————		——— 900-1000 ———	
High Idle (No-Load)	——— 2650-2700 ———		———2650-2700———	
Rated (Full Load)	——— 2500 ———		——— 2500 ———	
Rated Power at Pto	14.5 kW (19.5 hp)	17.5 kW (23.5 hp)	21.25 kW (28.5 hp)	25.75 kW (34.5 hp)
SIZES—CLEARANCES				
Crankshaft Main Journal Diameter	57.957-57.97 mm (2.281-2.282 in.)	67.951-67.97 mm (2.6750-2.6759 in.)	67.957-67.97 mm (2.6755-2.6759 in.)	
Main Bearing Radial Clearance	————— 0.044-0.116 mm ————— (0.0017-0.0045 in.)		——— 0.056-0.131 mm (0.002-0.005 in.)	
Crankshaft End Play	——— 0.10-0.40 mm ——— (0.004-0.016 in.)		——— 0.10-0.45 mm (0.004-0.018 in.)	
Crankpin Diameter	43.96-43.97 mm (1.730-1.731 in.)	——— 51.964-51.975 mm ——— (2.0458-0.0025 in.)		59.95-59.97 mm (2.0458-2.0463 in.)
Rod Bearing Radial Clearance	————— 0.035-0.083 mm ————— (0.001-0.003 in.)		——— 0.040-0.104 mm (0.002-0.004 in.)	

CONDENSED SERVICE DATA (CONT.)

		Models		
SIZES—CLEARANCES (Cont.)	1520	1720	1920	2120
Connecting Rod Side Clearance		0.10-0.30 mm (0.004-0.012 in.)		
Piston-to-Cylinder Bore Clearance	0.088-0.106 mm (0.0034-0.0041 in.)	0.038-0.064 mm (0.0015-0.0025 in.)	0.042-0.076 mm (0.0017-0.0030 in.)	0.087-0.139 mm (0.0034-0.0050 in.)

SPECIAL TORQUES				
Connecting Rod Caps		49-54 N·m (36-40 ft.-lbs.)		78-83 N·m (58-62 ft.-lbs.)
Crankshaft Main Bearing Holders		49-54 N·m (36-40 ft.-lbs.)		71-81 N·m (51-58 ft.-lbs.)
Main Bearing Holder Retaining Bolts		49-54 N·m (36-40 ft.-lbs.)		71-81 N·m (51-58 ft.-lbs.)
Crankshaft Pulley Retaining Nut		274-333 N·m (203-246 ft.-lbs.)		
Flywheel Bolts		59-69 N·m (44-51 ft.-lbs.)		
Cylinder Head Bolts		88-93 N·m (65-69 ft.-lbs.)		

CAPACITIES				
Cooling System	Note 2	5.6 liters (5.9 U.S. qts.)	5.6 liters (5.9 U.S. qts.)	8 liters (8.5 U.S. qts.)
Crankcase with Filter Change	4.5 liters (4.8 U.S. qts.)	4.5 liters (4.8 U.S. qts.)	6 liters (6.3 U.S. qts.)	8 liters (8.5 U.S. qts.)
Fuel Tank	27 liters (7.1 U.S. gals.)	32 liters (8.5 U.S. gals.)	39 liters (10.3 U.S. gals.)	42 liters (11.1 U.S. gals.)
Rear Axle and Transmission	22 liters (23.3 U.S. qts.)	27 liters (28.5 U.S. qts.)	29 liters (30.6 U.S. qts.)	33 liters (34.9 U.S. qts.)
Power Steering		1.8 liters (1.9 U.S. qts.)		
Front Wheel Drive Axle Housing	2.8 liters (3 U.S. qts.)	5 liters (5.3 U.S. qts.)	5 liters (5.3 U.S. qts.)	4.4 liters (4.6 U.S. qts.)

Note 2: Model 1520 cooling system capacity is 4 liters (4.2 U.S. qts.) with gear transmission and 5 liters (5.3 U.S. qts.) with hydrostatic transmission.

FRONT SYSTEM (Two Wheel Drive)

AXLE ASSEMBLY

Models 1120-1220

1. REMOVE AND REINSTALL. Models 1120 and 1220 may be equipped with either fixed tread width type axle or adjustable tread width axle.

To remove front axle assembly, support tractor behind the axle and remove front wheels. On models equipped with mechanical steering, disconnect steering drag link from steering arm. On models equipped with power steering, disconnect hoses from steering cylinder and plug all openings. On all models, remove nut (22—Fig. 1 or 2) from axle pivot shaft (24). Drive pivot shaft out of front axle and axle support, then lower axle assembly from tractor.

Inspect thrust washers (26), pivot bushings (27), pivot shaft (24) and axle main member (28) for excessive wear or damage. Clearance between pivot shaft and bushings should be 0.02-0.15 mm (0.001-0.006 inch) and wear limit is 0.3 mm (0.012 inch). Maximum allowable axle end play in support housing is 0.5 mm (0.020 inch).

To reinstall axle assembly, reverse the removal procedure. Install shims (25) as necessary to obtain desired axle end play of 0.3 mm (0.012 inch).

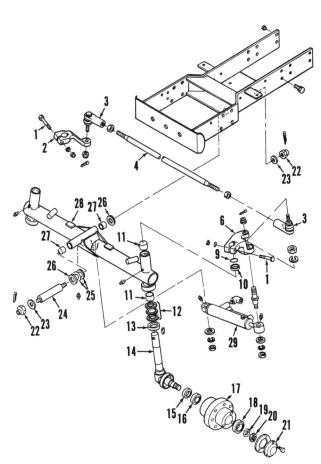

Fig. 1—Exploded view of fixed tread width front axle assembly used on Models 1120 and 1220 equipped with power steering.

1. Clamp bolt	18. Bearing
2. Steering arm	19. Lockwasher
3. Tie rod end	20. Nut
4. Tie rod	21. Hub cap
6. Steering arm	22. Nuts
9. "O" ring	23. Washers
10. Spacer	24. Pivot shaft
11. Bushings	25. Shims
12. Thrust bearing	26. Thrust washers
13. Seal	27. Bushings
14. Spindle	28. Axle main
15. Seal	member
16. Bearing	29. Power steering
17. Wheel hub	cylinder

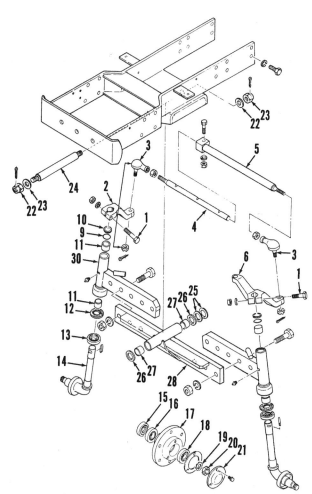

Fig. 2—Exploded view of adjustable tread width front axle assembly used on Models 1120 and 1220 equipped with mechanical steering. Refer to Fig. 1 for legend except for the following:

4. Tie rod link
5. Tie rod tube
30. Axle extension

Models 1320-1520-1720-1920-2120

2. REMOVE AND REINSTALL. Model 1320 is equipped with a fixed tread width front axle. An adjustable tread width front axle is standard equipment on Models 1520, 1720, 1920 and 2120.

To remove front axle assembly, support tractor behind the axle and remove front wheels. Disconnect hoses from power steering cylinder and plug all openings. Unbolt and remove axle pivot front bearing retainer (1—Fig. 3, 4 or 5). Remove retaining cap screws from rear bearing support (6), then move axle rearward from front bearing support and lower axle from tractor.

Inspect thrust washers (3), pivot bushings (4) and axle main member (10) for excessive wear or damage. Renew seals (5) if necessary. Clearance between axle pivot shafts and bushings should be 0.02-0.15 mm (0.001-0.006 inch) and wear limit is 0.30 mm (0.012 inch). Front axle end play in bearing supports should not exceed 0.20 mm (0.008 inch).

To reinstall axle assembly, reverse the removal procedure. Install shims (2) as necessary to obtain desired axle end play.

FRONT WHEEL BEARINGS

Models 1120-1220

3. REMOVE AND REINSTALL. It is recommended that front wheel bearings be removed, cleaned and repacked with grease after every 600 hours of operation.

To remove front wheel bearings, raise and support front of tractor. Remove wheel and tire. Remove hub cap (21—Fig. 1 or 2) and retaining nut (20). Withdraw wheel hub (17) and bearings from spindle. Remove wheel seal (15) and inner bearing (16) from hub.

Check bearings for pitting, roughness or other damage and renew as necessary. Renew wheel seals (15).

Pack wheel bearings with grease, then reinstall by reversing the removal procedure. Tighten wheel hub nut (20) to a torque of 30-35 N·m (22-26 ft.-lbs.), then loosen nut until tab of lockwasher can be bent into slot in nut. A slight drag should be felt when rotating the hub. Install wheel and tighten lug bolts to a torque of 58-73 N·m (43-54 ft.-lbs.).

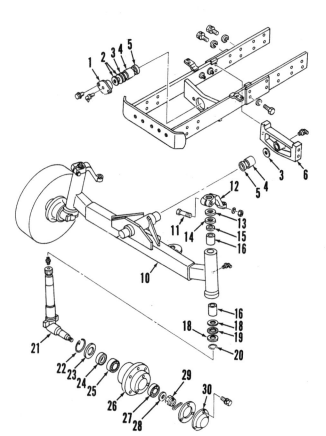

Fig. 3—Exploded view of fixed tread width front axle assembly used on Model 1320.

1. Pivot bearing retainer	16. Bushings
2. Shims	18. Thrust washers
3. Thrust washers	19. Needle bearing
4. Bushings	20. "O" ring
5. Seal	21. Spindle
6. Axle rear support	22. Retaining ring
10. Axle main member	23. Washer
	24. Seal
11. Clamp bolt	25. Bearing
12. Steering arm	26. Wheel hub
13. Shim	27. Bearing
14. Spacer	28. Washer
15. Seal	29. Nut
	30. Hub cap

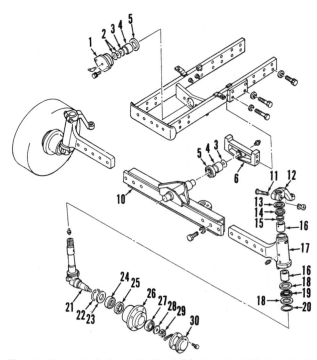

Fig. 4—Exploded view of adjustable tread width front axle assembly used on Model 1520. Refer to Fig. 3 for legend except for axle extension (17).

Models 1320-1520-1720-1920-2120

4. REMOVE AND REINSTALL. It is recommended that front wheel bearings be removed, cleaned and repacked with grease after every 600 hours of operation.

To remove front wheel bearings, raise and support front of tractor. Remove wheel and tire. Remove hub cap (30—Fig. 3, 4 or 5) and retaining nut (29). Withdraw wheel hub (26) and bearings from spindle. On Models 1320 and 1520, remove retaining ring (22) and washer (23). On all models, remove wheel seal (24) and inner bearing (25) from wheel hub.

Check bearings for pitting, roughness or other damage and renew as necessary. Renew wheel seals (24).

Pack wheel bearings with grease, then reinstall by reversing the removal procedure. Tighten wheel hub nut (29) while rotating hub until a drag is felt, then loosen nut to first castellation and install cotter pin.

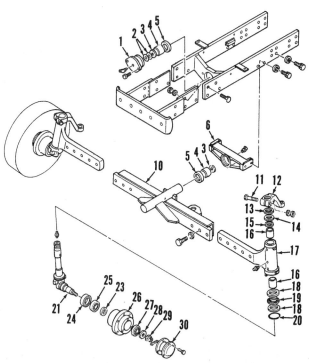

Fig. 5—Exploded view of adjustable tread width front axle assembly used on Model 1720. Axle assembly used on Models 1920 and 2120 is similar except that spacer (23) is not used on Model 2120.

1. Pivot bearing retainer
2. Shims
3. Thrust washers
4. Bushings
5. Seals
6. Axle rear support
10. Axle main member
11. Clamp bolt
12. Steering arm
13. Shims
14. Spacer
15. Seal
16. Bushings
17. Axle extension
18. Thrust washers
19. Needle bearing
20. "O" ring
21. Spindle
23. Spacer
24. Seal
25. Bearing
26. Wheel hub
27. Bearing
28. Washer
29. Nut
30. Hub cap

Install wheel and tighten lug bolts to a torque of 58-73 N·m (43-54 ft.-lbs.) on Model 1520, 66-83 N·m (48-61 ft.-lbs.) on Models 1720 and 1920 and 93-117 N·m (69-87 ft.-lbs.) on Model 2120.

SPINDLES AND BUSHINGS

Models 1120-1220

5. REMOVE AND REINSTALL. To remove spindles (14—Fig. 1 or 2), support front end of tractor with suitable stand and remove front wheels and wheel hub as outlined in paragraph 3. Remove clamp bolts (1) from steering arms (2 and 6), then tap spindles out of the steering arms and remove spindles from axle. Drive spindle bushings (11) out of axle if necessary.

Inspect all parts for excessive wear and renew as necessary. Use a suitable bushing driver to install new spindle bushings (11). Drive bushings in until they bottom against counterbore shoulder in axle. Renew seal (13) and "O" ring (9).

Lubricate spindle and bushings with grease, then reinstall spindle, steering arm and clamp bolts.

Models 1320-1520-1720-1920-2120

6. REMOVE AND REINSTALL. To remove spindles (21—Fig. 3, 4 or 5), raise and support front of tractor. Remove front wheels and wheel hub as outlined in paragraph 4. Remove clamp bolt (11) from steering arms (12), then tap spindle downward out of steering arm and axle. Drive bushings (16) out of axle if necessary.

Inspect all parts for excessive wear and renew as necessary. Use a suitable bushing driver to install new spindle bushings (16). Upper bushing should be recessed (R—Fig. 6) below upper surface of axle as follows: 5 mm (0.197 inch) on Models 1320 and 1520; 7 mm (0.275 inch) on Models 1720 and 1920; 8 mm (0.315 inch) on Model 2120. Lower bushing should be bottomed against counterbore shoulder. Renew "O" ring (20) and seal (15).

Lubricate spindle and bushings with grease. Reinstall spindle and steering arm, using shims (13) as necessary to remove end play from spindle. Install clamp bolt (11) and tighten securely.

TIE RODS AND TOE-IN

All Models

7. Nonadjustable automotive type tie rod ends are used on all tractors. Tie rod ends must be renewed if excessively worn.

Recommended front wheel toe-in is 0-5 mm (0-3/16 inch), measured at front and rear of wheels at wheel spindle height. To check toe-in, mark front of the wheels at hub height (Fig. 7) and measure the distance between the marks. Roll tractor forward until marks are positioned at the rear at hub height and

again measure distance between the marks. The measurement at the front should be 0-5 mm (0-3/16 inch) less than measurement at the rear.

To adjust toe-in, loosen locknuts (1) on tie rod ends. On Models 1320, 1520, 1720, 1920 and 2120, disconnect power steering cylinder (2) from tie rod (3). On all models, turn tie rod as necessary to obtain recommended toe-in. After toe-in is correct, tighten locknuts.

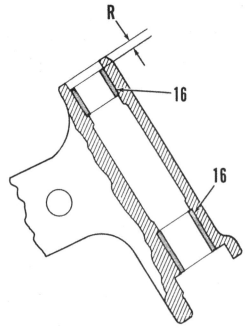

Fig. 6—When installing new spindle bushings on Models 1320, 1520, 1720, 1920 and 2120, drive in upper bushing (16) until specified recession (R) is obtained. Refer to text.

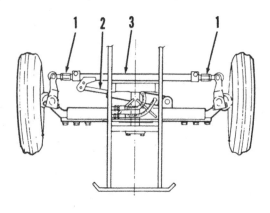

Toe-In Check Marks

Fig. 7—To check front wheel toe-in, measure distance between wheels at front and rear at spindle height. Refer to text.

FRONT WHEEL DRIVE AXLE

LUBRICATION

All Models

8. Recommended lubricant for front wheel drive axle is Ford 134 lubricant or equivalent. Oil level in front wheel drive axle housing should be maintained between the mark on dipstick (Fig. 10) and lower end of dipstick. Axle housing oil capacity is approximately 1.8 liters (1.9 U.S. quarts) on Models 1120 and 1220, 2.8 liters (3 U.S. quarts) on Models 1320 and 1520, 5 liters (5.2 U.S. quarts) on Models 1720 and 1920, 4.4 liters (4.6 U.S. quarts) on Model 2120. Oil level in final drive housings should be maintained at oil level/fill plug opening (Fig. 10). Manufacturer recommends changing oil in axle housing and final drive housings every 300 hours of operation. Drain plugs are located in bottom of final drive housings and differential center housing.

Fig. 10—View of front wheel drive axle showing location of axle housing oil level dipstick and final drive housing drain plug and filler/oil level plug.

R&R AXLE ASSEMBLY

All Models

9. To remove axle assembly, first raise front of tractor and place suitable supports behind the axle. Remove front wheels and drive shaft. On Models 1120 and 1220 with mechanical steering, disconnect steering drag link from steering arm. On all models equipped with power steering, disconnect hydraulic hoses from steering cylinder and plug all openings. Support axle housing with a floor jack. Remove retaining bolts from front pivot bracket (1—Fig. 11 or 12) and rear pivot bracket (10), then lower axle assembly from the tractor.

Inspect axle pivot bushings (2 and 9) for wear or damage. Clearance between axle housing trunnions and pivot bushings should be 0.02-0.19 mm (0.001-0.007 inch) at the front and 0.02-0.16 mm (0.001-0.006 inch) at the rear. Renew bushings if clearance exceeds 0.35 mm (0.014 inch) at front or rear. Use a suitable bushing driver to install new bushings. Be sure that bushings are recessed into axle pivot brackets far enough to allow for installation of seals (3 and 8) in the pivot brackets.

To reinstall axle assembly, reverse the removal procedure. Check axle housing fore and aft end play. Desired end play is 0.30 mm (0.012 inch) or less. On Models 1120 and 1220, shims (4—Fig. 11) are available for adjusting axle end play. On all other models, end play is adjusted by loosening rear bearing holder mounting bolts and sliding bearing holder forward or rearward as necessary to obtain desired end play. Lubricate pivot bushings with multipurpose lithium base grease.

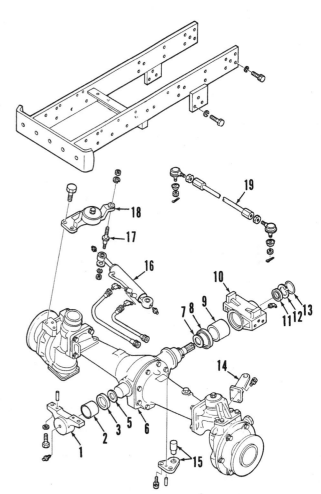

Fig. 11—View of front wheel drive axle used on Models 1120 and 1220.

1. Front pivot support	6. Axle housing
2. Bushing	7. Thrust washer
3. Seal	8. Seal
4. Shims	9. Bushing
5. Thrust washer	10. Rear pivot housing

Fig. 12—View of front wheel drive axle used on Models 1720, 1920 and 2120. Axle used on Models 1320 and 1520 is similar except that seal (11), snap ring (12) and "O" ring (13) are not used.

1. Front pivot bracket	8. Seal
2. Bushing	9. Bushing
3. Seal	10. Rear pivot housing
5. Thrust washer	11. Seal
6. Axle housing	12. Snap ring
7. Thrust washer	13. "O" ring

FINAL DRIVE AND WHEEL HUB

All Models

10. R&R AND OVERHAUL. To remove final drive assembly from either side, first support front of tractor and remove front wheel and tire. Drain oil from outer drive housing and axle center housing. Disconnect drag link (mechanical steering) or power steering cylinder and tie rod from steering arm. Support final drive assembly with a suitable hoist, then remove cap screws attaching pinion gear case (15—Fig. 13 or 14) to axle housing (30) and withdraw final drive assembly.

It is suggested that backlash between bevel gear (6) and pinion gear (10) be checked prior to disassembly to determine if gears or bearings are excessively worn. To check backlash, remove oil drain plug from reduction gear housing (8) and thread a long bolt (3—Fig. 15) into oil drain hole until it contacts pinion gear to prevent gear from moving. Install a bolt (2) in wheel hub flange and position a dial indicator (1) against head of bolt. Rotate wheel hub back and forth and note dial indicator reading. Normal backlash is 0.10-0.25 mm (0.004-0.010 inch). If backlash exceeds 0.50 mm (0.020 inch), renew bearings and/or gears as necessary.

To disassemble final drive, remove retaining cap screws from outer cover (3—Fig. 13 or 14) and withdraw cover with wheel hub (1), bevel gear (6) and

bearings as an assembly. Pull bevel gear and inner bearing (7) off wheel hub shaft. Remove retaining ring (5), then drive wheel hub shaft out of bearing (4) and cover (3). Drive oil seal (2) from cover.

Remove cap screws attaching steering arm (23) to reduction gear housing (8). Separate reduction gear housing (8) from pinion gear case (15). On Models 1320, 1520, 1720, 1920 and 2120, remove bottom cover (12—Fig. 14), bearing (11) and pinion gear (10). On all models, remove drive shaft (9—Fig. 13 or 14), kingpin (18), bearings and pinion gears from the housings.

Inspect all parts for excessive wear or damage and renew as necessary. Renew "O" rings and seals when reassembling. On Models 1320 and 1520, adjust kingpin bearings to zero end play and zero preload by means of shims (22—Fig. 14). On Models 1720, 1920 and 2120, adjust kingpin bearings to zero end play and zero preload by turning adjusting screw (34) as necessary.

DIFFERENTIAL

All Models

11. R&R AND OVERHAUL. The differential assembly can be removed from axle housing without removing axle housing from tractor if desired. Drain oil from axle housing. Raise and support front of tractor, then remove left front wheel. On Models 1120 and 1220, disconnect steering drag link (mechanical steer-

1. Wheel hub & shaft
2. Seal
3. Cover
4. Bearing
5. Snap ring
6. Bevel drive gear
7. Bearing
8. Bevel gear housing
9. Drive shaft
10. Pinion gear
11. Bearing
13. Bearing
14. Seal
15. Pinion gear housing
16. Bearing
17. Snap ring
18. Kingpin
19. "O" ring
20. Bearing
23. Steering arm
24. "O" ring
25. Pinion gear
28. Pinion gear
29. Bearing
30. Axle housing, R.H.

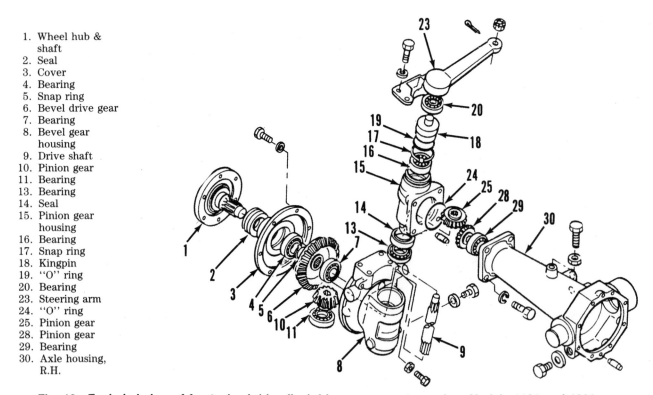

Fig. 13—Exploded view of front wheel drive final drive components used on Models 1120 and 1220.

ing) or power steering cylinder from left steering arm. On all models, disconnect tie rod from left steering arm. Unbolt and remove left axle housing and final drive unit as an assembly from right axle housing. Remove differential and ring gear from right axle housing.

To disassemble differential on Models 1120 and 1220, remove carrier bearings and side gears (20—Fig. 16) from differential case (30). Drive out retaining pin (27) and remove pinion gear shaft (26), pinion gears (23) and thrust washers (22).

To disassemble differential on Models 1320, 1520, 1720, 1920 and 2120, unbolt and remove differential case cover (18—Fig. 17 or 18), side gear (20) and thrust washer (19). On Models 1320, 1520, 1720 and 1920, remove pinion shaft retaining ring (29). On all models, slide pinion shafts (25 and 26) out of differential case and remove pinion gears (24), thrust washers (22) and pinion shaft support (21). Remove the other side gear (20) and thrust washer (19).

Inspect all parts for excessive wear or damage and renew as necessary. Refer to the following specifications:

Models 1120-1220

Side gear to pinion gear backlash
  Standard . . . . .0.10-0.15 mm
  (0.004-0.006 in.)
  Wear limit . . . . .0.5 mm
  (0.020 in.)

Pinion shaft to pinion gear
radial clearance
  Standard . . . . .0.1-0.3 mm
  (0.004-0.012 in.)
  Wear limit . . . . .0.5 mm
  (0.020 in.)

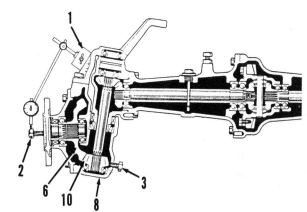

Fig. 15—Check bevel gear (6) to pinion gear (10) backlash using a dial indicator (1) as shown. Refer to text.

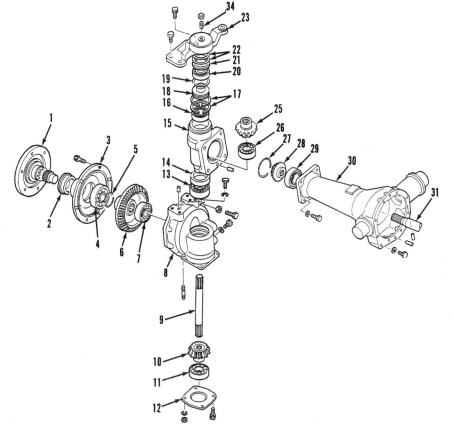

1. Wheel hub & shaft
2. Seal
3. Cover
4. Bearing
5. Split retaining ring
6. Bevel drive gear
7. Bearing
8. Bevel gear housing
9. Drive shaft
10. Pinion gear
11. Bearing
12. Cover
13. Bearing
14. Seal
15. Pinion gear housing
16. Bearing
17. Snap rings
18. Kingpin
19. Snap ring
20. Bearing
21. Spacer
22. Shims
23. Steering arm
25. Pinion gear
26. Bearing
27. Retaining ring
28. Pinion gear
29. Bearing
30. Axle housing, R.H.
31. Axle shaft
34. Adjusting screw

Fig. 14—Exploded view of front wheel drive final drive components typical of Models 1320, 1520, 1720, 1920 and 2120. Shims (22) are used on Models 1320 and 1520. Spacer (21) and adjusting screw (34) are used on Models 1720, 1920 and 2120.

Pinion gear thrust washer thickness

Standard . . . . .1.1-1.2 mm
(0.043-0.047 in.)

Wear limit . . . . .1.0 mm
(0.039 in.)

Models 1320-1520

Side gear to pinion gear backlash

Standard . . . . .0.10-0.15 mm
(0.004-0.006 in.)

Wear limit . . . . .0.3 mm
(0.012 in.)

Pinion shaft to pinion gear
radial clearance

Standard . . . . .0.1 mm
(0.004 in.)

Wear limit . . . . .0.5 mm
(0.020 in.)

Pinion gear thrust washer thickness

Standard . . . . .1.2 mm
(0.047 in.)

Wear limit . . . . .0.8 mm
(0.032 in.)

Models 1720-1920-2120

Side gear to pinion gear backlash (1720-1920)

Standard . . . . .0.10-0.15 mm
(0.004-0.006 in.)

Wear limit . . . . .0.5 mm
(0.020 in.)

Side gear to pinion gear backlash (2120)

Standard . . . . .0.15-0.20 mm
(0.006-0.008 in.)

Wear limit . . . . .0.25 mm
(0.010 in.)

Pinion shaft to pinion gear
radial clearance

Standard . . . . .0.1 mm
(0.004 in.)

Wear limit . . . . .0.5 mm
(0.020 in.)

Pinion gear thrust washer thickness

Standard . . . . .1.2 mm
(0.047 in.)

Wear limit . . . . .0.8 mm
(0.032 in.)

Side gear thrust washer thickness

Wear limit . . . . .0.9 mm
(0.035 in.)

To reassemble differential, reverse the disassembly procedure. If ring gear (28) was removed, renew ring gear retaining bolts and use locking plates (31) to prevent bolts from loosening. On Model 2120, be sure that holes in pinion shafts (25 and 26—Fig. 18) are aligned with holes for dowel bolts (27). On all models, bevel pinion to ring gear backlash and differential carrier bearing preload must be adjusted as outlined in paragraphs 13 and 14 if any of the following components were renewed: Differential assembly com-

4. Axle housing, R.H.
5. Bevel pinion shaft
6. Thrust washer
7. Shims
8. Bearings
10. "O" ring
11. Collar
12. Seal
13. Locknut
15. Shims
16. Bearings
17. Spacer
20. Differential side gears
22. Thrust washer
23. Differential pinion gears
26. Pinion gear shaft
27. Roll pin
28. Bevel ring gear
30. Differential case
31. Locking plate
32. Differential carrier bearing
33. Spacer
34. Axle bearing
35. Shims
36. Axle shaft
37. Axle housing, L.H.

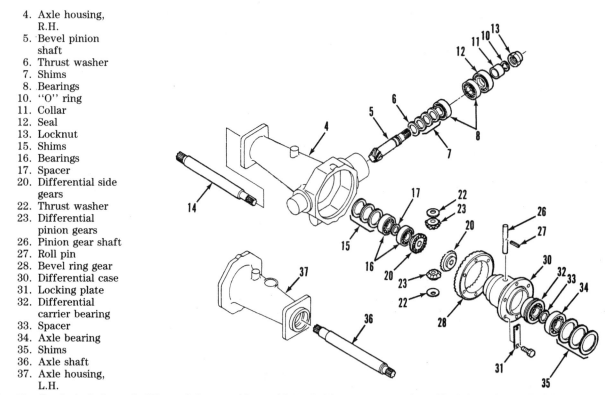

Fig. 16—Exploded view of differential assembly and bevel drive gears used on Models 1120 and 1220.

ponents, differential carrier bearings, axle shaft inner bearings, axle shafts or axle housings. If none of these components are being renewed, reassemble differential and front axle using original shims (15 and 35) in their original locations.

BEVEL DRIVE GEARS

All Models

12. R&R AND OVERHAUL. The bevel pinion (5—Fig. 16, 17 or 18) and ring gear (28) must be renewed as a matched set.

To remove bevel drive gears, drain oil from axle housing. Remove axle assembly from tractor as outlined in paragraph 9. Disconnect tie rod from left steering arm. Separate left axle housing from right axle housing and remove differential and ring gear assembly from right axle housing. Remove ring gear

retaining bolts and separate ring gear from differential case.

Remove retaining nut (13) from bevel pinion shaft (5), then tap shaft inward and remove from axle housing. Drive oil seal (12), collar (11) and rear bearing (8) from axle housing. Remove front bearing (8), shims (7) and thrust washer (6), if used, from bevel pinion shaft. Be sure to retain shims (7) for use in reassembly.

To reassemble, install original shims (7), thrust washer and front bearing onto bevel pinion shaft. Insert shaft into axle housing, then install rear bearing, "O" ring and oil seal onto rear of shaft. Lubricate "O" ring and oil seal with clean oil, then install collar (11) and locknut (13). Adjust pinion bearing preload as follows:

Wrap a cord around pinion shaft as shown in Fig. 20, then use a spring scale to measure pull required to rotate pinion shaft. Tighten locknut to obtain following spring scale reading:

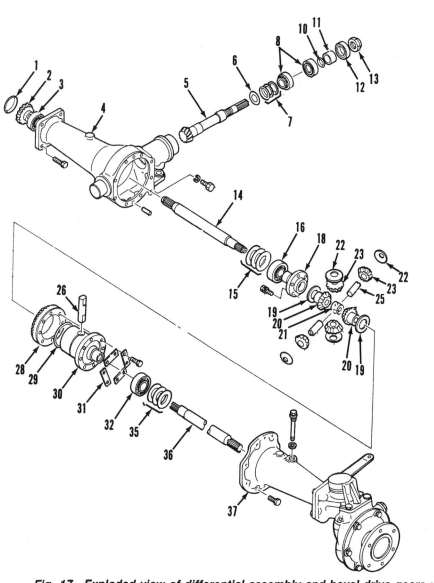

1. "O" ring
2. Pinion gear
3. Bearing
4. Axle housing, R.H.
5. Bevel pinion shaft
6. Thrust washer
7. Shims
8. Bearings
10. "O" ring
11. Collar
12. Oil seal
13. Locknut
14. Axle shaft
15. Shims
16. Differential carrier bearing
18. Differential case cover
19. Thrust washer
20. Side gears
21. Pinion shaft support
22. Thrust washer
23. Differential pinion gears
25. Pinion gear shafts (2)
26. Pinion gear shaft
28. Bevel ring gear
29. Retaining ring
30. Differential case
31. Locking plates
32. Differential carrier bearing
35. Shims
36. Axle shaft
37. Axle housing, L.H.

Fig. 17—Exploded view of differential assembly and bevel drive gears used on Models 1320 and 1520.

Models 1120-1220

New bearings . . . . .8-12 kg
(18-26 lbs.)

Used bearings . . . . .4-6 kg
(9-13 lbs.)

Models 1320-1520

New bearings . . . . .13-17 kg
(29-37 lbs.)

Used Bearings . . . . .6.5-8.5 kg
(14.5-18.5 lbs.)

Models 1720-1920

New bearings . . . . .14.5-17.5 kg
(32-39 lbs.)

Used bearings . . . . .7-9 kg
(16-20 lbs.)

Model 2120

New bearings . . . . .12-15 kg
(27-33 lbs.)

Used bearings . . . . .6-8 kg
(14-17 lbs.)

When bearing preload is correctly set, install lock-washer (13A—Fig. 18) and second locknut (13) on Models 1720 and 1920. Tighten locknut and bend tabs of lockwasher to prevent loosening. On all other models, stake locknut (13—Fig. 16, 17 or 18) at shaft groove to prevent loosening.

NOTE: During factory assembly, ring gear retaining bolts are coated with a thread locking compound and locking plates (31) are not used. When installing new ring gear, discard factory installed bolts and replace with new bolts and locking plates.

If differential case (30), carrier bearings (16 and 32), axle bearings, axle shafts or axle housings were renewed, differential carrier bearing preload must be adjusted as outlined in paragraph 14. Bevel ring gear to pinion backlash and gear mesh should be adjusted

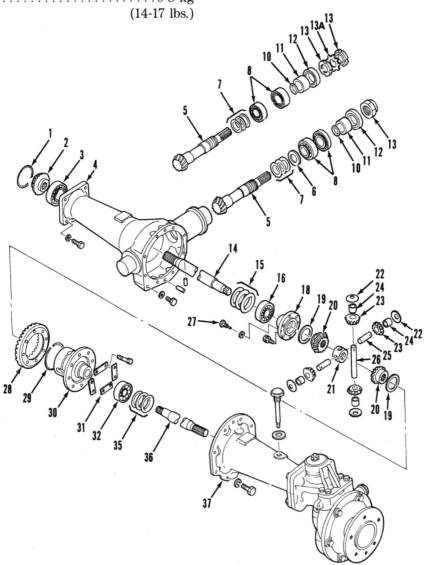

Fig. 18—Exploded view of differential assembly and bevel drive gears used on Models 1720, 1920 and 2120. Two locknuts (13) and lockwasher (13A) are used on Models 1720 and 1920. Single locknut (13) and spacer (6) are used on Model 2120. Differential pinion shafts (25 and 26) are held in place by retaining ring (29) on Models 1720 and 1920 and by special dowel bolts (27) on Model 2120. Refer to Fig. 17 for legend except for bushings (24).

as outlined in paragraph 13. If none of these components are being renewed, reinstall differential and ring gear assembly using original shims (15 and 35) in their original locations.

13. BEVEL PINION TO RING GEAR BACKLASH ADJUSTMENT. Bevel pinion to ring gear backlash is adjusted by means of shims (15—Fig. 16, 17 or 18) located in right axle housing (4). To adjust backlash, position the assembled right axle housing complete with axle shaft in a vertical position, supported by the axle housing and not the axle shaft. Install original shims in right axle housing, then position differential and ring gear assembly with carrier bearing in right axle housing. Make sure that carrier bearing is fully seated in axle housing. Place a dial indicator on axle housing as shown in Fig. 21, then rock ring gear back and forth while holding pinion shaft (5) and note dial indicator reading. Backlash should be 0.15-0.20 mm (0.006-0.008 inch) for Model 2120 and 0.10-0.15 mm (0.004-0.006 inch) for all other models. If backlash is not within specified limits, remove differential assembly and add or remove shims (15—

Fig. 16, 17 or 18) as necessary to obtain correct backlash.

After backlash is correctly set, adjust differential carrier bearings as outlined in paragraph 14.

14. DIFFERENTIAL CARRIER BEARING ADJUSTMENT. After backlash between bevel pinion and ring gear is correctly set as outlined in paragraph 13, adjust pinion bearings as follows: Position assembled right axle housing complete with differential assembly in a vertical position. Install more shims (35—Fig. 16, 17 or 18) than will be required in left axle housing (37). Position left axle housing over differential assembly, making sure that left carrier bearing is fully seated in axle housing.

NOTE: There should be a gap between the two axle housings. If not, install additional shims.

Install four bolts equally spaced around left axle housing and tighten finger tight. Be sure that gap between the housings is equal on all sides. Using a feeler gage (G—Fig. 22), measure the gap between the housings. Remove left axle housing and remove shims (35—Fig. 16, 17 or 18) equal to the measured gap. This will provide specified zero end play and zero preload adjustment of carrier bearings.

As a final assembly check, apply Prussian Blue to teeth of ring gear. Install left axle housing and tighten retaining bolts securely. Rotate bevel pinion shaft until ring gear has rotated one full turn. Remove left

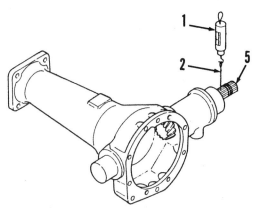

Fig. 20—To adjust bevel pinion bearing preload, use a spring scale (1) and cord (2) to measure pounds of pull required to rotate pinion shaft (5). Tighten pinion shaft nut to obtain specified preload. Refer to text.

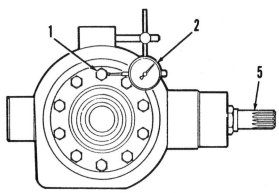

Fig. 21—Use a dial indicator (2) to measure backlash between bevel ring gear (1) and pinion (5).

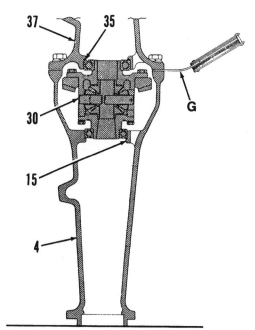

Fig. 22—Differential carrier bearings (15 and 35) should be adjusted by means of shims (35) to obtain zero end play and zero preload. Refer to text for adjustment procedure.

axle housing and check gear tooth contact pattern using Fig. 23 as a reference. If necessary, adjust ring gear and pinion to obtain correct tooth pattern.

TRANSFER CASE

All Models

15. R&R AND OVERHAUL. To remove front wheel drive transfer case, first drain oil from rear axle center housing and transfer case. Remove drive shaft cover and drive shaft. Unbolt and remove transfer case assembly from bottom of axle center housing.

To disassemble, drive out roll pin (10—Fig. 25) and withdraw idler gear shaft (11), idler gear (9), thrust washers (7) and needle bearing (8). Unbolt and remove seal housing (2) from gear housing (12). Withdraw drive shaft (6) with front bearing (5) from gear housing, then remove drive gear (13) and rear bearing (14).

Inspect all parts for excessive wear or damage and renew as necessary. Renew all "O" rings (1 and 3) and oil seal (4).

To reassemble, reverse the disassembly procedure. Lubricate "O" rings and lip of oil seal prior to installation and be careful not to damage seal lip during installation. Position thrust washers (7) with dimpled side facing the idler gear (9).

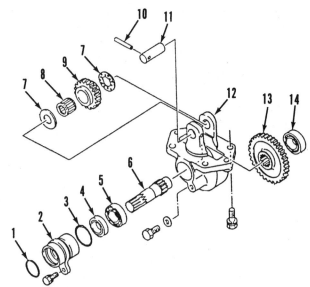

Fig. 25—Exploded view of front wheel drive transfer case assembly typical of all models.

1. "O" ring	8. Needle bearing
2. Seal housing	9. Idler gear
3. "O" ring	10. Roll pin
4. Oil seal	11. Gear shaft
5. Bearing	12. Gear housing
6. Drive shaft	13. Drive gear
7. Thrust washers	14. Bearing

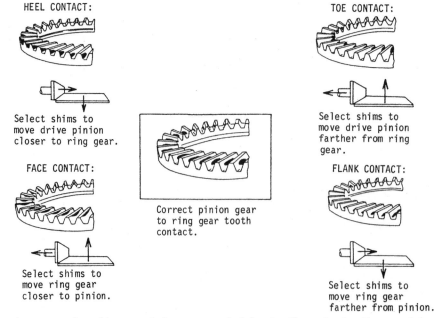

Fig. 23—Drawings showing examples of incorrect ring gear and pinion tooth contact pattern and desired tooth contact pattern (center drawing).

MANUAL STEERING GEAR

Models 1120-1220

16. All manual steering models are equipped with a recirculating ball nut and sector shaft type steering gear, which is mounted on top of clutch housing.

Steering wheel free play can be externally adjusted. Desired free play (F—Fig. 26) of steering wheel is 20 to 30 mm (3/4 to 1-1/8 inches), measured at outer rim of steering wheel. If free play is excessive, loosen locknut (2) and turn adjusting screw (1) clockwise to reduce free play. Tighten locknut to secure adjusting screw after desired setting is obtained.

17. REMOVE AND REINSTALL. To remove steering gear, remove steering wheel cap and retaining nut. Remove steering wheel using a suitable puller. Remove rear cover from instrument panel shroud. Disconnect instrument panel wiring harness connectors. Disconnect tachometer cable from tachometer. Disconnect hand throttle cable from injection pump. Remove knobs from hand throttle lever and main shift lever or hydrostatic speed control lever. Remove screws attaching fuel gage to instrument panel. Remove screws attaching instrument panel shroud to engine baffle plate, then remove instrument panel and shroud assembly.

Clamp fuel supply line to prevent fuel leakage, or drain fuel from tank. Disconnect fuel line from fuel filter and disconnect fuel return hoses from top of fuel tank. Remove fuel tank retaining strap and remove fuel tank from tractor. Remove fuel tank support plate. Remove bolts retaining shift linkage bracket (3—Fig. 27) or hydrostatic speed control bracket to steering column housing (2).

Scribe reference marks across steering gear sector shaft (5) and pitman arm to ensure correct reassembly, then remove pitman arm using a suitable puller. Remove steering gear retaining cap screws and remove steering gear assembly (4) from clutch housing.

To reinstall steering gear, reverse the removal procedure. Be sure to align reference marks on sector shaft and pitman arm made prior to removal. Bleed air from fuel system as outlined in paragraph 53.

18. OVERHAUL. To disassemble removed steering gear, proceed as follows: Remove retaining screws from side cover (3—Fig. 28), then withdraw cover and sector shaft (7) as an assembly. Remove locknut (1) and turn adjusting screw (6) to remove sector shaft from side cover. Remove screws attaching steering column cover (28) to housing (9) and remove cover, steering shaft (22), worm shaft (19) and ball nut (18) as an assembly. Separate steering column from worm shaft and ball nut assembly. Note that worm shaft and ball nut are available only as an assembly and should not be disassembled.

Inspect all parts for wear or damage. Renew worm shaft and ball nut assembly if worm shaft binds or does not turn smoothly. Renew sector shaft (7), side cover (3) and/or bushing (10) if clearance between

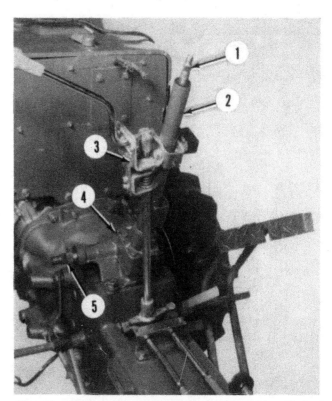

Fig. 27—View of manual steering gear assembly with fuel tank, instrument panel and steering column shroud removed.

1. Steering shaft
2. Column housing
3. Gear shift bracket
4. Steering gear assy.
5. Sector shaft

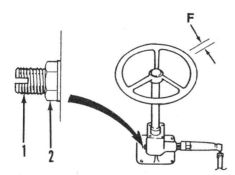

Fig. 26—Steering wheel free travel (F) should be 20 to 30 mm (3/4 to 1-1/8 inches). Turn steering gear adjusting screw (1) clockwise to reduce free travel.

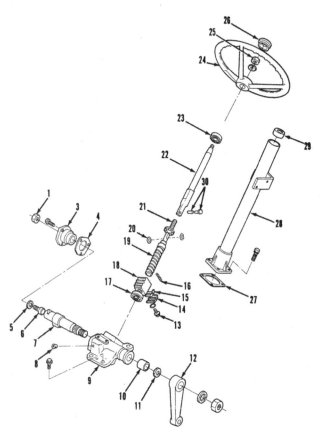

Fig. 28—Exploded view of manual steering gear assembly that is standard equipment on Models 1120 and 1220.

1. Locknut
3. End cover
4. Gasket
5. Shim
6. Adjusting screw
7. Sector gear & shaft
8. Oil level/filler plug
9. Gear housing
10. Bushing
11. Oil seal
12. Pitman arm
13. Screw
14. Retainer
15. Ball guides
16. Balls
17. Bearing
18. Ball nut
19. Worm shaft
20. "O" rings
21. Spring
22. Steering shaft
23. Bearing
24. Steering wheel
25. Retaining nut
26. Cap
27. Shim
28. Column cover
29. Bushing
30. Pins

shaft and bushings is excessive. Renew gasket (4) and oil seal (11).

When reassembling, install ball nut, steering shaft and column cover using original shims (27) between steering column and gear housing. Check and adjust steering shaft bearing preload as follows: Attach a cord to steering shaft and measure pull required to rotate shaft using a spring scale as shown in Fig. 29. Bearing preload is correct when a pull of 2-4.5 kg (4.5-9.9 lbs.) is required to rotate shaft. Adjust preload as necessary by adding or removing shims (27—Fig. 28).

Assemble sector shaft (7), adjusting bolt (6) and shim (5) in side cover (3). Locate ball nut (18) in center of travel on worm shaft (19), then install sector gear and shaft with center tooth of sector gear engaging center teeth of ball nut. Tighten side cover retaining cap screws.

Fill steering gear to level plug (8) opening with Ford 134 hydraulic fluid or equivalent. Temporarily install steering wheel and check steering wheel free play. Turn adjusting bolt as necessary to obtain specified free play of 20 to 30 mm (3/4 to 1-1/8 inches). Install pitman arm (12) aligning reference marks on arm and sector shaft made prior to disassembly.

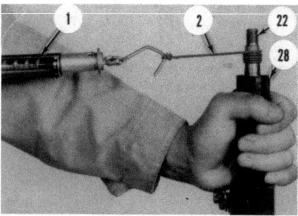

Fig. 29—To check steering shaft bearing preload, wrap a cord (2) around shaft (22) and use a spring scale (1) to measure pull required to rotate shaft.

POWER STEERING SYSTEM

Hydrostatic power steering is optional on Models 1120 and 1220 and standard on Models 1320, 1520, 1720, 1920 and 2120. The power steering system consists of a hydraulic pump and reservoir, steering control valve and steering cylinder. Pressure oil is directed from the pump to steering hand control valve where it is either directed to the steering cylinder when turning the steering wheel or returned to the reservoir when valve is in neutral position. The steering valve unit will also act as a manual oil pump when it is not being supplied oil from the hydraulic pump for any reason, such as when the engine is stopped.

LUBRICATION AND BLEEDING

All Models So Equipped

19. The power steering oil reservoir is located above the power steering hydraulic pump, which is mounted on the engine timing gear case. A dipstick (Fig. 31) is provided for checking the oil level. Oil level should be checked with tractor on a level surface and with engine shut off.

Power steering oil should be changed after every 600 hours of operation. A drain plug is located in bottom of the oil reservoir. Recommended oil is Ford 134 hydraulic fluid. Capacity is approximately 1.8 liters (1.9 U.S. quarts) for all models.

To bleed air from steering system, start engine and run at about 1500 rpm. Turn steering wheel from lock to lock until steering feels smooth. This will purge air from system.

Fig. 31—Power steering reservoir and hydraulic pump are located on front of the engine.

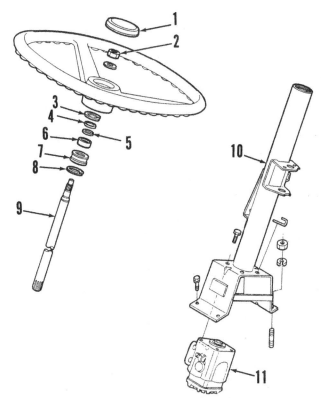

Fig. 32—Exploded view of steering column used on models with power steering.

1. Cap	7. Rubber seal
2. Nut	8. Snap ring
3. Snap ring	9. Steering shaft
4. Spacer	10. Steering column
5. Snap ring	11. Power steering
6. Bearing	control valve

TROUBLE-SHOOTING

All Models So Equipped

20. The following table lists some of the troubles which may occur in the operation of power steering system and their possible causes.

1. Hard or no steering movement.
   a. Low oil level.
   b. Excessive load on front wheels.
   c. Steering pressure relief valve faulty.
   d. Faulty steering control valve unit.
   e. Hydraulic pump defective.
2. Steering wheel turns on its own or will not return to neutral.
   a. Steering control valve spool binding.
3. Steering cylinder will not follow steering wheel rotation.
   a. Air in steering system.
   b. Steering cylinder piston seal faulty.
   c. Steering control valve improperly assembled, resulting in valve mistiming.
4. Steering wheel "kicks back."
   a. Hydraulic pipes improperly connected to steering control valve.
   b. Steering control valve improperly assembled, resulting in valve mistiming.

TESTING

All Models

21. The steering pressure relief valve (1—Fig. 33) is located in the steering control valve body. Relief valve is not adjustable and is not available separately. If relief valve is faulty, steering control valve must be renewed.

To check steering relief pressure, disconnect one of the hydraulic lines to steering cylinder and connect 0-20000 kPa (0-3000 psi) pressure gage to the line using a "Tee" fitting. Start engine and operate at 1500 rpm. Turn and hold steering wheel against lock position and observe pressure gage reading. Specified relief pressure is 11755 kPa (1705 psi) for Model 2120 and 9790-10480 kPa (1420-1520 psi) for all other models. If pressure is low, it indicates that either the relief valve, hydraulic pump or steering cylinder is faulty.

To check steering cylinder for leakage past the piston seal, disconnect hydraulic line from end of cylinder opposite the test gage and plug the cylinder port. Repeat steering relief valve pressure test. If pressure is now within specified limits, steering cylinder is faulty.

To check hydraulic pump output, connect a suitable hydraulic flow meter to pressure outlet of pump.

Position flow meter return line in oil filler opening of reservoir.

> CAUTION: With flow meter connected to hydraulic pump in this manner, there is no relief valve in the system to protect the pump. Make certain that flow meter restrictor valve is fully open before starting engine, and do not exceed specified testing pressure.

Start engine and run at 2500 rpm. Record pump flow at zero pressure. Carefully adjust flow meter restrictor valve until pump pressure has risen to 11720 kPa (1700 psi) on Model 2120 or 10000 kPa (1450 psi) on all other models and record pump flow. Specified flow rate is 5.6 lpm (1.5 gpm) for Models 1120, 1220, 1320 and 1520; 8.7 lpm (2.3 gpm) for Model 1720; 16 lpm (4.2 gpm) for Model 1920; 22 lpm (5.8 gpm) for Model 2120. Oil flow under pressure should be within 15 percent of oil flow at zero pressure; if not, pump is faulty.

STEERING CONTROL VALVE

All Models So Equipped

22. REMOVE AND REINSTALL. To remove steering control valve assembly, first disconnect battery ground cable. Remove steering wheel cap (1—Fig. 32) and retaining nut (2), then remove steering wheel using a suitable puller. Remove snap ring (3) and withdraw steering shaft (9) from top of steering column (10). Remove dash rear panel and disconnect key switch wiring connector. Remove steering column side shrouds. Drain oil from power steering reservoir, then disconnect power steering tubes at both ends and move tubes away from steering valve. On Model 2120, disconnect hydraulic shuttle shift pressure tube (if so equipped). On all models, remove control valve mounting bolts and remove control valve from tractor.

To reinstall, reverse the removal procedure. Refill reservoir with clean oil. Start engine and turn steering from stop to stop several times to bleed air from system.

23. OVERHAUL. To disassemble control valve unit, mount valve body in a soft jawed vise with end cover (20—Fig. 33) facing upward. Do not overtighten the vise as valve body can be easily distorted and damaged. Remove end cover retaining bolts and remove end cover, gerotor pump (18), drive shaft (17) and spacer plate (16).

> NOTE: Lapped surfaces on end cover, gerotor set, spacer plate and valve body must be protected from scratches, burrs or other damage as sealing of these parts depends upon their finish and flatness.

Remove valve body from vise and place on a clean lint-free shop towel. On Models 1920 and 2120, unbolt and remove mounting plate from top of valve body. On Models 1120, 1220, 1320, 1520 and 1720, remove retaining ring (3) from top of valve body. Rotate spool (12) and sleeve (10) so pin (11) is positioned horizontally. Push on bottom end of spool and sleeve to force seal gland (4) out of valve body. Remove spool and sleeve as an assembly from valve body. To separate spool from sleeve, remove pin (11) and centering springs (13 and 14) from spool. While rotating spool to prevent binding, slide spool out bottom of sleeve. If necessary, remove pressure relief valve assembly (1).

Thoroughly clean all parts in a suitable solvent and air dry. Inspect all parts for excessive wear, scoring, scratches or other damage and renew as necessary. Individual parts, other than "O" rings and oil seals, are not available for control valve unit used on Models 1920 and 2120.

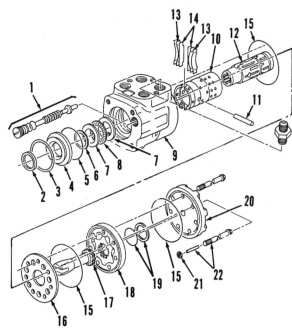

Fig. 33—Exploded view of power steering control valve assembly used on Models 1120, 1220, 1320, 1520 and 1720. Control valve unit used on Models 1920 and 2120 is similar.

1. Pressure relief valve assy.
2. Dust seal
3. Retaining ring
4. Seal gland
5. "O" ring
6. Oil seal
7. Thrust races
8. Needle bearing
9. Valve body
10. Sleeve
11. Drive pin
12. Valve spool
13. Centering springs (curved)
14. Springs (flat)
15. "O" rings
16. Spacer plate
17. Drive shaft
18. Gerotor pump
19. "O" ring & seal
20. End cover
21. Check ball
22. Retainer bolt & pin

To reassemble, install relief valve assembly making sure that valve plug does not protrude above the surface of valve body. Coat spool and sleeve with clean hydraulic oil, then insert spool into sleeve. Be sure that spool rotates freely in the sleeve. Align centering spring slots in spool and sleeve and insert special spring installing tool (Fig. 34) through the slots. Position ends of springs in groove of installing tool as shown in Fig. 34. Be sure that notched edges of springs face downward. While pinching outer end of springs together, slide tool out of the spool leaving springs positioned in slots. Install pin (11—Fig. 33) through spool and sleeve, then insert spool and sleeve assembly into valve body from the bottom. Make certain that sleeve rotates freely in valve body bore.

On Models 1120, 1220, 1320, 1520 and 1720, install "O" ring (5—Fig. 33) in valve body. Position needle bearing (8) and thrust races (7) on the spool. Install dust seal (2) and oil seal (6) in seal gland (4) as shown in Fig. 35. Install seal gland and retaining ring (3—Fig. 33) in valve body. Position valve body in vise with bottom end facing upward. Install check ball (21—Fig. 36) into retaining bolt bore.

On Models 1920 and 2120, install needle bearing (8) and thrust races (7) with chamfered edges of thrust races facing away from the bearing as shown in Fig. 37. Install new seal in mounting plate, then install

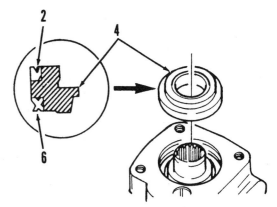

Fig. 35—Seal gland (4) is used on Models 1120, 1220, 1320, 1520 and 1720. Lip of dust seal (2) should face outward and lip of oil seal (6) should face inward.

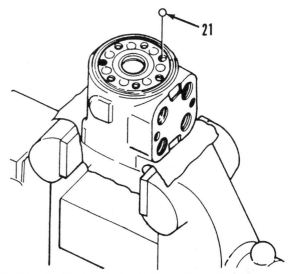

Fig. 36—On Models 1120, 1220, 1320, 1520 and 1720, be sure that check ball (21) is positioned in retainer bolt hole.

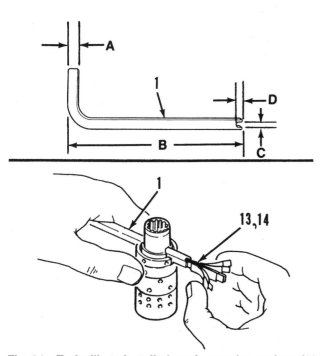

Fig. 34—To facilitate installation of centering springs (13 and 14), it is recommended that installation tool (1) similar to one shown be used. Tool may be fabricated from 6 mm (1/4 inch) square keystock.

A. 6 mm (1/4 inch)
B. 150 mm (6 inches)
C. 3.5 mm (9/64 inch)
D. 6 mm (1/4 inch)

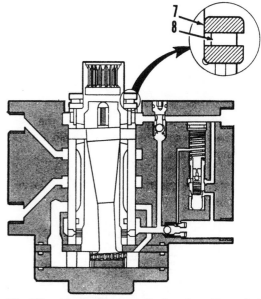

Fig. 37—When installing needle bearing (8) and thrust races (7), position races with chamfered edges facing away from needle bearing.

mounting plate on valve body. Position valve body in vise with bottom end facing upward.

On all models, position "O" ring (15—Fig. 33) and spacer plate (16) on valve body. Turn spool and sleeve assembly in valve body so that drive pin (11) is positioned parallel with port face (P) of valve body as shown in Fig. 38. Scribe a reference mark on end of drive shaft (17) parallel with slot for drive pin, then position drive shaft in valve body engaging slot with drive pin. Position stator and rotor assembly (18) on valve body, aligning reference mark on end of drive shaft with two of the valleys (2—Fig. 39) of pump rotor. When correctly installed and timed, one rotor lobe will be centered on stator lobe as shown in Fig. 39.

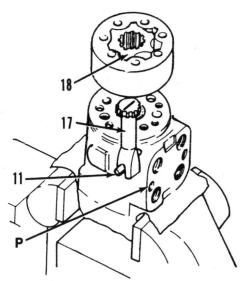

Fig. 38—Drive pin (11) and two of the valleys of gerotor pump rotor (18) must be positioned parallel to port face (P) of valve body to correctly "time" the control valve assembly.

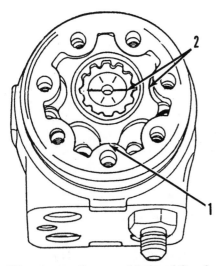

Fig. 39—When correctly assembled and timed, one rotor lobe will be centered on one stator lobe (1) as shown.

Install "O" ring and seal (19—Fig. 33) in the pump rotor. Install end cover (20) with new "O" ring (15). Install end cover retaining bolts, making sure that bolt with pin (22) is positioned in same hole as check ball (21). Tighten retaining bolts in steps to a torque of 28 N·m (21 ft.-lbs.) following tightening sequence shown in Fig. 40. Be sure that valve spool turns freely without binding.

POWER STEERING PUMP

All Models So Equipped

24. R&R AND OVERHAUL. To remove power steering hydraulic pump (3—Fig. 42 or 43), first remove drain plug and drain oil from reservoir (1). Disconnect oil suction tube (2) and pressure tube (4) from pump. Remove pump mounting stud nuts (6) and withdraw pump from engine.

> NOTE: The only parts available for servicing the hydraulic pump are shaft seal and internal "O" rings. Complete pump assembly must be renewed if pump is faulty.

Refer to Fig. 44 for exploded view of hydraulic pump used on Models 1120, 1220, 1320 and 1520 or to Fig. 45 for exploded view of pump assembly used on Models 1720, 1920 and 2120. To disassemble pump, remove cap screws retaining end cover (10) to pump housing and carefully separate end cover, bearings (6—Fig. 44) or wear plates (6—Fig. 45) and pump gears from the housing.

> NOTE: Identify all parts as they are disassembled so that each part can be installed in its original position when reassembling pump.

Inspect parts for excessive wear, scratches or other damage and renew pump assembly if necessary.

Lubricate all parts with clean hydraulic oil during reassembly. On Models 1720, 1920 and 2120 (Fig. 45), note that gear wear plates (6) are not the same. Be

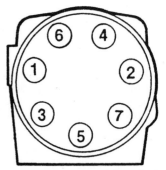

Fig. 40—Tighten end cover retaining bolts in steps following sequence shown.

sure that wear plate with two holes in it is installed on end cover side of gears, and that rectangular slots in wear plate bores are positioned on pressure side of pump.

To reinstall pump, reverse the removal procedure. Refill reservoir with oil, start engine and turn steering wheel from lock to lock several times to bleed air from steering system.

POWER STEERING CYLINDER

All Models So Equipped

25. OVERHAUL. To disassemble steering cylinder, straighten cylinder barrel crimping from cylinder gland head (12—Fig. 46). Unscrew cylinder head from cylinder barrel, then pull piston rod (15) and piston (5) from cylinder. Remove retaining nut (3) and withdraw piston and cylinder gland head from piston rod. Remove "O" rings and back-up rings from piston. Remove "O" rings, back-up ring (if used), oil seal and wiper ring from cylinder gland head.

Inspect cylinder barrel, piston and rod for scuffing, score marks or other damage. Clearance between piston and cylinder barrel should not exceed 0.7 mm (0.028 inch). Clearance between piston rod and cylinder head bushing (11) should not exceed 0.3 mm (0.012 inch). Clearance between anchor pins and

Fig. 43—On Model 2120, power steering hydraulic pump (3) is located on left side of engine.

1. Power steering reservoir	4. Pressure line
2. Suction line	5. Steering cylinder oil lines
3. Power steering pump	8. Hydraulic shift shuttle pump

Fig. 42—Power steering hydraulic pump (3) is located on right side of engine on all tractors except Model 2120.

1. Power steering reservoir	6. Pump retaining nuts
2. Suction line	7. Pressure line
3. Power steering pump	8. Hydraulic system pump
4. Pressure line	9. Suction line
5. Steering cylinder oil lines	10. Steering oil return line

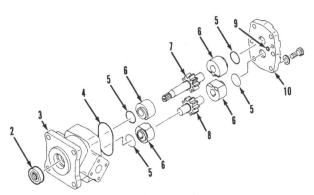

Fig. 44—Exploded view of power steering hydraulic pump used on Models 1120, 1220, 1320 and 1520.

2. Shaft seal	
3. Pump housing	7. Drive shaft & gear
4. Seal ring	8. Idler shaft
5. "O" rings	9. "O" ring
6. Shaft bearings	10. End cover

bearings or bushings (1 and 16) should not exceed 0.5 mm (0.019 inch). Renew all "O" rings, back-up rings, oil seal and wiper ring.

Lubricate all parts prior to reassembly. Tighten piston retaining nut to a torque of 35 N·m (26 ft.-lbs.) on Models 1120, 1220, 1320 and 1520 or 46 N·m (34 ft.-lbs.) on Models 1720, 1920 and 2120. Stake the nut in two places to prevent it from loosening. Tighten cylinder gland head to a torque of 127 N·m (94 ft.-lbs.) on Models 1120, 1220, 1320 and 1520 or 167 N·m (123 ft.-lbs.) on Models 1720, 1920 and 2120. Stake the cylinder barrel at gland head notch to prevent gland head from loosening.

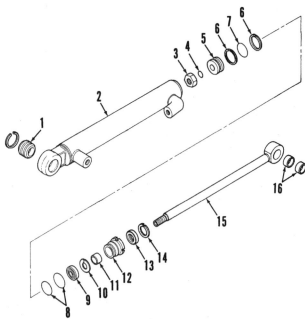

Fig. 46—Exploded view of power steering cylinder used on Models 1320 and 1520 with two wheel drive. Steering cylinder used on other models is similar except that bushings are used in place of spherical bearing (1) on tractors equipped with front wheel drive axle and back-up ring (10) is not used on Models 1720, 1920 and 2120.

1. Spherical bearing	10. Back-up ring
2. Cylinder barrel	11. Bushing
3. Nut	12. Cylinder gland
4. "O" ring	head
5. Piston	13. Wiper ring
6. Back-up rings	14. Snap ring
7. "O" ring	15. Piston rod
8. "O" rings	16. Bushings
9. Oil seal	

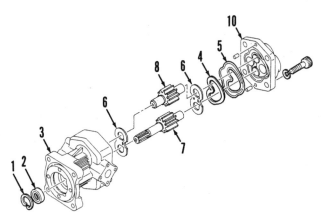

Fig. 45—Exploded view of power steering hydraulic pump used on Models 1720, 1920 and 2120.

1. Snap ring	6. Wear plates
2. Shaft seal	7. Drive shaft &
3. Pump housing	gear
4. Back-up ring	8. Idler gear
5. Seal ring	10. End cover

ENGINE AND COMPONENTS

R&R ENGINE ASSEMBLY

All Models

27. To remove engine, first drain coolant from radiator, drain hydraulic oil from transmission housing and drain oil from power steering reservoir.

Separate front end assembly from engine as follows: Remove front end weights if so equipped. Disconnect battery cables. Disconnect headlight wiring, then remove hood. Remove engine side screens and lower panels. Disconnect upper and lower radiator hoses, coolant overflow hose and radiator and engine block drain hoses (4—Fig. 48). Remove radiator support brace (1). Disconnect battery cable from starter solenoid (7). If equipped with manual steering, disconnect drag link from steering gear. If equipped with power steering, disconnect power steering tubes (2) from side frame. If equipped with hydrostatic transmission, remove oil pipes from oil cooler located

in front of radiator. If equipped with front wheel drive axle, disconnect the drive shaft. Raise front of tractor and position safety stand under clutch housing. Support front end assembly with suitable overhead hoist. Remove cap screws attaching side rails (6) to engine, then carefully roll front end assembly away from engine.

Separate engine from clutch housing as follows: Remove the steering wheel. Remove steering column shroud center panel. Disconnect wiring connectors from instrument panel and release wiring harness from retaining clips. Remove throttle control cables (8—Fig. 49) from injection pump. Disconnect glow plug wire, coolant temperature sender unit wire, oil pressure sender wire, air cleaner sender wire and fuel stop solenoid wire. On Model 2120, remove fuel stop solenoid assembly (10). On all models, disconnect tachometer cable from engine. Unbolt and remove instrument panel and steering column shrouds. Disconnect fuel inlet line at fuel shut-off valve (9) and plug fuel line or drain fuel from tank. Disconnect fuel return lines from fuel tank. Remove fuel tank retaining band and remove fuel tank from tractor. Remove starter motor. If equipped with power steering, remove hydraulic tubes from steering con-

Fig. 48—View of left side of engine on Model 2120. Power steering lines (2), coolant drain hoses (4) and power steering pump (5) are located on right side of tractor on all other models.

1. Radiator brace
2. Power steering lines
3. Hydraulic shift shuttle pump (2120 only)
4. Coolant drain hoses
5. Power steering pump
6. Side rails
7. Starter motor

Fig. 49—View of right side of engine on Model 2120.

1. Radiator brace
8. Throttle control cables
9. Fuel shut-off valve
10. Fuel stop solenoid (2120)
11. Hydraulic pressure tube
12. Hydraulic suction tube
13. Lower radiator hose
14. Upper radiator hose

trol valve and steering pump. Remove hydraulic tubes from hydraulic system pump and hydraulic shift shuttle transmission pump if so equipped. Plug all openings. Remove air cleaner assembly and muffler. Support engine with overhead hoist. Remove cap screws attaching engine to clutch housing and carefully separate engine from clutch housing.

To reinstall engine, reverse the removal procedure.

CYLINDER HEAD

All Models

28. To remove cylinder head, first drain coolant from radiator and cylinder block. Remove air cleaner assembly. Disconnect radiator hoses and remove coolant overflow tank if so equipped. On Models 1120 and 1220, remove the radiator, water pump and thermostat housing. On all other models, remove thermostat housing. On models with power steering, drain oil from steering system reservoir and remove reservoir and mounting bracket. On all models except 2120, disconnect oil transfer tube from front of cylinder head. On Model 1920, remove steering wheel, instrument panel, steering column side and top shrouds, fuel tank and front baffle plate to provide clearance for removal of rocker arm housing. On all models, remove muffler and exhaust manifold. Remove injector pressure lines and leak-off lines, then remove the injectors. Remove glow plugs. Disconnect wires from oil pressure sender and coolant temperature sender. Remove intake manifold and rocker cover. Unbolt and remove rocker arm assemblies and push rods. Loosen cylinder head bolts gradually in reverse order of tightening sequence shown in Figs. 50, 51 and 52 to prevent warping of cylinder head, then remove head from engine.

Using a valve spring compressor, remove valve spring retainer locks, valve springs and valves. Keep valve components separated and identified for reassembly in their original locations if reused. On Models 1120 and 1220, tap prechambers out of fuel injector bores in cylinder head.

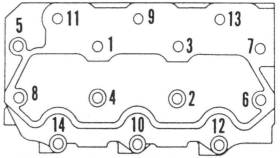

Fig. 50—Cylinder head bolt tightening sequence for Models 1120 and 1220.

NOTE: Glow plugs must be removed prior to removing injector prechambers.

Thoroughly clean the cylinder head, then check for cracks or other damage. Use a straightedge and feeler gage to check cylinder head for flatness as shown in Fig. 53. Resurface or renew cylinder head if warped more than 0.12 mm (0.005 inch).

Head gaskets of different thicknesses are available for service, and selection of correct gasket thickness is based on the distance pistons protrude above face of cylinder block with pistons at top dead center. Use a dial indicator or other suitable means to measure height of each piston above cylinder block surface as shown in Fig. 54. Use the measurement from piston that has the highest protrusion and select appropriate thickness head gasket as indicated in chart shown in Fig. 55.

NOTE: The variation in amount of protrusion between all pistons must not exceed 0.1 mm (0.004 inch).

The last four digits of head gasket part number are stamped on the gasket. Install head gasket with part number digits or ''TOP'' marking facing up. Lubricate threads of cylinder head retaining bolts with engine oil. Tighten head bolts in three steps following sequence shown in appropriate Fig. 50, 51 or 52 to specified final torque of 52-56 N·m (38-41 ft.-lbs.) on Models 1120 and 1220 or 88-93 N·m (65-69 ft.-lbs.) on Models 1320, 1520, 1720, 1920 and 2120.

Install rocker arm assembly and adjust valve clearance as outlined in paragraph 29. Complete installation by reversing the removal procedure. Bleed air from fuel system as outlined in paragraph 53.

VALVE CLEARANCE ADJUSTMENT

All Models

29. Valve clearance should be checked after every 600 hours of operation and whenever cylinder head has been removed. Clearance should be checked with engine cold and with piston at top dead center on compression stroke (both valves closed). To check clearance, remove rocker arm cover and

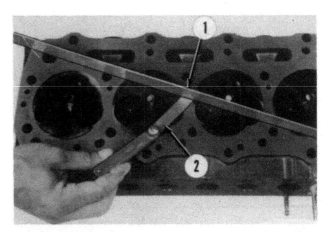

Fig. 53—A straightedge (1) and feeler gage (2) can be used to check cylinder head gasket surface for flatness as shown.

Fig. 51—Cylinder head bolt tightening sequence for Models 1320, 1520 and 1720.

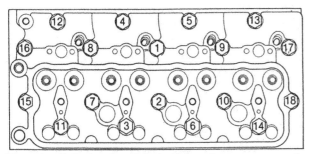

Fig. 52—Cylinder head bolt tightening sequence for Models 1920 and 2120.

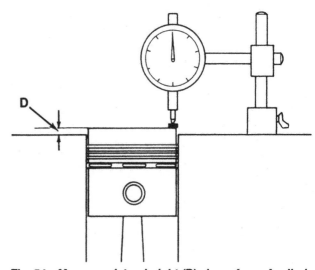

Fig. 54—Measure piston height (D) above face of cylinder block to determine proper thickness cylinder head gasket to install. Refer to text and Fig. 55.

turn crankshaft to position number one piston at TDC. Use a feeler gage to check clearance between rocker arm and valve stem (Fig. 56). On Model 2120, specified clearance is 0.30 mm (0.012 inch) for intake and exhaust. On all other models, specified clearance is 0.20 mm (0.008 inch) for intake and exhaust. To adjust clearance, loosen locknut and turn rocker arm adjusting screw as necessary. Rotate crankshaft to bring each successive piston in firing order to TDC and check valve clearance. Firing order for Models 1920 and 2120 is 1-3-4-2 and for all other models is 1-2-3.

ROCKER ARMS AND PUSH RODS

All Models

30. On Models 1120 and 1220, a single shaft (2—Fig. 58) mounted on top of cylinder head supports the rocker arms (4). A cap screw (6) at each end of rocker shaft retains the rocker arms on the shaft, and a spring pin (1) locates shaft in front rocker support (3). On Models 1320, 1520, 1720 and 1920, a single support shaft (2—Fig. 59) and the rocker arms (4) are mounted in a separate support casting (3) which is

Tractor Model	Piston Height Measurement	Gasket Thickness	Part Number SBA –
1120	0.55-0.65 mm (0.022-0.026 in.) 0.65-0.75 mm (0.026-0.030 in.) 0.75-0.85 mm (0.030-0.033 in.)	1.1 mm (0.043 in.) 1.2 mm (0.047 in.) 1.3 mm (0.051 in.)	111146821 111146831 111146840
1220	0.45-0.55 mm (0.018-0.022 in.) 0.55-0.65 mm (0.022-0.026 in.) 0.65-0.75 mm (0.026-0.030 in.)	1.1 mm (0.043 in.) 1.2 mm (0.047 in.) 1.3 mm (0.051 in.)	111146882 111146892 111146901
1320	0.50-0.60 mm (0.020-0.024 in.) 0.60-0.70 mm (0.024-0.028 in.) 0.70-0.80 mm (0.028-0.032 in.)	1.2 mm (0.047 in.) 1.3 mm (0.051 in.) 1.4 mm (0.055 in.)	111147150 111147160 111147170
1520, 1720	0.50-0.60 mm (0.020-0.024 in.) 0.60-0.70 mm (0.024-0.028 in.) 0.70-0.80 mm (0.028-0.032 in.)	1.2 mm (0.047 in.) 1.3 mm (0.051 in.) 1.4 mm (0.055 in.)	111146981 111146991 111147001
1920	0.50-0.60 mm (0.020-0.024 in.) 0.60-0.70 mm (0.024-0.028 in.) 0.70-0.80 mm (0.028-0.032 in.)	1.2 mm (0.047 in.) 1.3 mm (0.051 in.) 1.4 mm (0.055 in.)	111147180 111147190 111147200
2120	Under 0.60 mm (0.024 in.) Over 0.60 mm (0.024 in.)	1.5 mm (0.059 in.) 1.7 mm (0.067 in.)	111147210 111147220

Fig. 55—Cylinder head gaskets are available in different thicknesses to provide correct piston to cylinder head clearance.

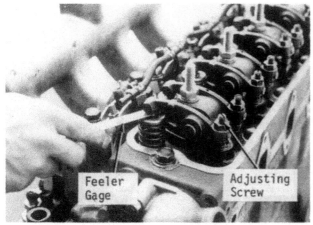

Fig. 56—Use a feeler gage to check valve clearance with engine cold and piston at TDC on compression stroke.

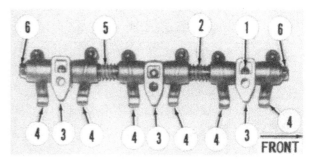

Fig. 58—Valve rocker assembly used on Models 1120 and 1220.

1. Spring pin
2. Rocker shaft
3. Shaft supports
4. Rocker arms
5. Spring
6. Retaining screws

bolted to top of cylinder head. A set screw (1) locates rocker shaft in the support casting. On Model 2120, individual support bracket and shaft assemblies (3—Fig. 60) are used to separately support rocker arms of each cylinder. Rocker arms (4) are retained on support shaft by snap rings (6). On all models, intake and exhaust rocker arms are not interchangeable. Note that rocker arms and push rods should be reinstalled in their original location if being reused.

Inspect all parts for wear or damage. Check push rods for straightness by rolling on a flat surface. Renew push rods if bent; do not attempt to straighten push rods. Check ends of push rods for excessive wear. If push rods are worn, inspect corresponding cam follower and rocker arm for excessive wear. Clearance between rocker shaft and bore of rocker

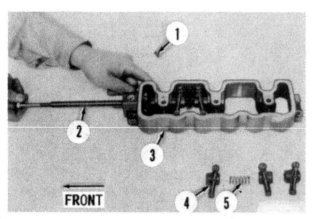

Fig. 59—Valve rocker assembly used on Models 1320, 1520, 1720 and 1920.

1. Set screw
2. Rocker shaft
3. Support casting
4. Rocker arms
5. Spring

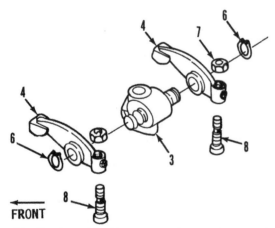

Fig. 60—On Model 2120, individual support bracket and shaft assemblies (3) are used to separately support rocker arms for each cylinder.

3. Rocker support & shaft assy.
4. Rocker arms
6. Snap rings
7. Locknut
8. Adjusting screws

arms should not exceed 0.20 mm (0.008 inch). Measure shaft OD and rocker arm ID and renew parts that do not meet the following wear limits.

Models 1120-1220-1320-1520-1720
Shaft OD
    Standard . . . . 11.65-11.67 mm
        (0.4587-0.4594 in.)
    Wear limit . . . . 11.57 mm
        (0.456 in.)
Shaft to rocker clearance
    Standard . . . . 0.032-0.068 mm
        (0.0013-0.0027 in.)

Model 1920
Shaft OD
    Standard . . . . 14.95-14.97 mm
        (0.5886-0.5894 in.)
    Wear limit . . . . 14.87 mm
        (0.5854 in.)
Shaft to rocker clearance
    Standard . . . . 0.032-0.068 mm
        (0.0013-0.0027 in.)

Model 2120
Shaft OD
    Standard . . . . 13.68-13.71 mm
        (0.539-0.540 in.)
    Wear limit . . . . 13.56 mm
        (0.534 in.)
Shaft to rocker clearance
    Standard . . . . 0.02-0.056 mm
        (0.001-0.002 in.)

VALVE SPRINGS

All Models

31. Valve springs are interchangeable for intake and exhaust valves. Springs should be renewed if discolored, distorted or if they fail to meet the following specifications:

Model 2120
Out of square (max.) . . . . 1.2 mm
        (0.047 in.)
Free length
    Standard . . . . 46 mm
        (1.811 in.)
    Minimum . . . . 44 mm
        (1.732 in.)
Test load . . . . 129 N
        (29 lbs.)
Test length . . . . 30.4 mm
        (1.197 in.)

All Other Models
Out of square (max.) . . . . 2.0 mm
        (0.079 in.)

Free length

Standard . . . . .35 mm

(1.378 in.)

Minimum . . . . .33.5 mm

(1.319 in.)

Test load . . . . .69-80 N

(15.5-17.8 lbs.)

Test length . . . . .30.4 mm

(1.197 in.)

VALVES, GUIDES AND SEATS

All Models

32. Valve face angle and seat angle is 45 degrees for intake and exhaust. Renew valves that are burned, bent, badly pitted or excessively worn. After refacing valves, check thickness of valve head margin (M—Fig. 61) and renew valves if margin is less than 1.0 mm (0.040 inch) on Model 2120 or 0.5 mm (0.020 inch) on all other models.

After refacing valve seat, apply Prussian Blue to valve seat or mark valve face with a soft lead pencil. Install valve and rotate valve against seat while applying light pressure against valve. Seat contact point should be in the center of valve face and seat width (S—Fig. 61) should be within specified limits. If necessary, use a 30 degree stone to lower seat contact point and use a 60 degree stone to raise seat contact point. Both stones will narrow the seat width. Check for correct recession (R) of valve head below machined surface of cylinder head. If valve recession exceeds specified wear limit, renew valve and/or seat or cylinder head. Renewable valve seat inserts are available on Models 1320, 1520, 1720 and 1920.

Measure valve stem diameter and valve guide bore to determine stem to guide clearance. Renew valves and/or cylinder head if clearance is excessive.

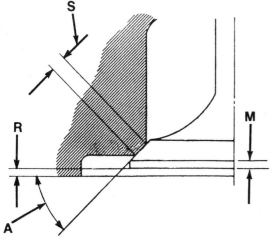

Fig. 61—Drawing of valve and seat showing 45 degree seat angle (A), valve head margin (M), valve recession (R) and seat width (S). Refer to text for specifications.

Valve Seat Width

Models 1120-1220

Standard . . . . .1.7-2.1 mm

(0.067-0.082 in.)

Maximum . . . . .2.5 mm

(0.098 in.)

Models 1320-1520-1720

Standard . . . . .1.6-1.8 mm

(0.063-0.071 in.)

Maximum . . . . .2.5 mm

(0.098 in.)

Model 1920

Standard . . . . .1.9-2.1 mm

(0.075-0.083 in.)

Maximum . . . . .2.8 mm

(0.110 in.)

Model 2120

Standard . . . . .1.7-2.5 mm

(0.067-0.098 in.)

Maximum . . . . .3.0 mm

(0.118 in.)

Valve Recession

Models 1120-1220-1320-1520-1720

Standard . . . . .0.85-1.15 mm

(0.0335-0.0453 in.)

Maximum . . . . .1.8 mm

(0.071 in.)

Model 1920

Standard . . . . .0.7-1.0 mm

(0.028-0.039 in.)

Maximum . . . . .1.6 mm

(0.063 in.)

Model 2120

Standard . . . . .1.0-1.2 mm

(0.039-0.047 in.)

Maximum . . . . .2.0 mm

(0.079 in.)

Valve Stem Diameter-Intake

Models 1120-1220-1320-1520-1720-1920

Standard .6.955-6.97 mm

(0.2738-0.2744 in.)

Minimum . . . . .6.89 mm

(0.271 in.)

Model 2120

Standard . . . . .7.95-7.97 mm

(0.313-0.314 in.)

Minimum . . . . .7.88 mm

(0.310 in.)

Valve Stem Diameter-Exhaust

Models 1120-1220

Standard . . . . .6.94-6.96 mm

(0.273-0.274 in.)

Minimum . . . . .6.84 mm

(0.269 in.)

Models 1320-1520-1720-1920

Standard . . . . .6.94-6.95 mm
(0.273-0.2736 in.)
Minimum .6.84 mm
(0.269 in.)

Model 2120

Standard .7.93-7.94 mm
(0.312-0.3125 in.)
Minimum .7.85 mm
(0.309 in.)

Stem to Guide Clearance-Intake
All Models

Standard .0.03-0.06 mm
(0.001-0.002 in.)
Maximum .0.20 mm
(0.008 in.)

Stem to Guide Clearance-Exhaust
Models 1120-1320-1520-2120

Standard . . . . .0.04-0.065 mm
(0.0016-0.0025 in.)
Maximum .0.25 mm
(0.010 in.)

Models 1220-1720-1920

Standard . . . . .0.05-0.075 mm
(0.002-0.003 in.)
Maximum .0.25 mm
(0.010 in.)

TIMING GEAR CASE

Models 1120-1220-1320-1520-1720-1920

33. REMOVE AND REINSTALL. To remove timing gear case (12—Fig. 62), first drain cooling system and engine oil. Disconnect headlight wiring connectors, then remove hood, engine side screens, lower panels, grille and grille side panels. Remove battery and battery support tray. Disconnect radiator hoses, remove radiator retaining nuts and support braces, and remove radiator from tractor. Remove engine fan and pulley. Remove alternator and mounting brackets. On Models 1320, 1520, 1720 and 1920, remove the water pump.

If equipped with power steering, drain oil from power steering reservoir tank. Disconnect hydraulic tubes from steering pump, then remove pump mounting nuts and remove power steering pump from front of timing gear case. If equipped with manual steering, remove hydraulic system pump (2) mounting bolts.

Disconnect throttle control cables (1) from governor lever. Remove injector lines and cap all openings. Remove fuel stop solenoid (4). Remove injection pump mounting bolts and raise injection pump (3) sufficiently to remove retaining pin (6—Fig. 63) and

disconnect governor control link (8) from injection pump control rack pin. Remove crankshaft pulley using a suitable puller. Unbolt and remove timing gear case from engine.

Inspect governor linkage and timing gear case for wear or damage and renew as necessary. Use a suitable driver to install new crankshaft oil seal (16—Fig. 63) in the case. Lubricate seal lip before installing timing gear case.

To reinstall timing gear case, reverse the removal procedure. Be sure to align hole in oil pump cover (6—Fig. 64) with pin (11) in timing gear case. Tighten crankshaft pulley nut to a torque of 48-58 N·m (36-43 ft.-lbs.) on Models 1120 and 1220, 274-333 N·m (203-246 ft.-lbs.) on Models 1320, 1520, 1720 and 1920. Bleed air from fuel system as outlined in paragraph 53.

Model 2120

34. REMOVE AND REINSTALL. To remove timing gear case (4—Fig. 65), first drain cooling system, engine oil, transmission/hydraulic system oil and power steering system oil. Disconnect headlight wiring connectors, then remove hood, engine side screens, lower panels, grille and grille side panels. Remove battery and battery support tray. Disconnect radiator hoses, remove radiator retaining nuts and support braces, and remove radiator from tractor. Remove engine fan and pulley. Remove alternator and mounting bracket as an assembly.

Remove support clamp attaching hydraulic system pressure tube and inlet tube to bottom of transmis-

Fig. 62—View of right side of engine on Model 1120 equipped with manual steering.

1. Throttle control cables	3. Injection pump
	4. Fuel stop solenoid
2. Hydraulic system pump	12. Timing gear case

sion housing. Disconnect pressure tube and inlet tube from hydraulic system pump and move inlet tube away from the engine. Disconnect hydraulic tubes

from power steering pump, then remove pump mounting bolts and remove power steering pump. If equipped with hydraulic shift shuttle transmission,

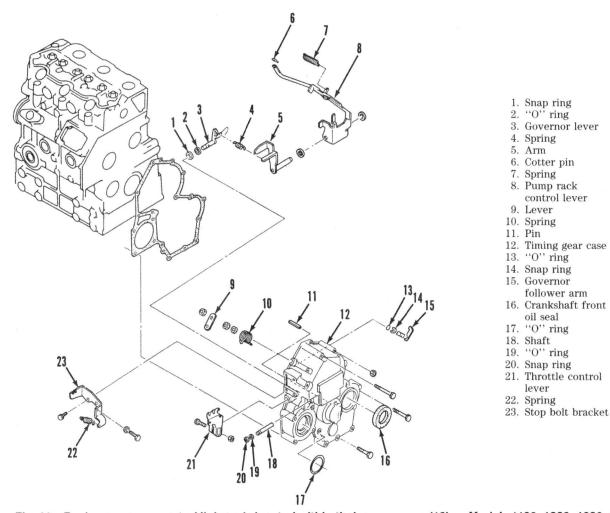

1. Snap ring
2. "O" ring
3. Governor lever
4. Spring
5. Arm
6. Cotter pin
7. Spring
8. Pump rack control lever
9. Lever
10. Spring
11. Pin
12. Timing gear case
13. "O" ring
14. Snap ring
15. Governor follower arm
16. Crankshaft front oil seal
17. "O" ring
18. Shaft
19. "O" ring
20. Snap ring
21. Throttle control lever
22. Spring
23. Stop bolt bracket

Fig. 63—Engine governor control linkage is located within timing gear case (12) on Models 1120, 1220, 1320, 1520, 1720 and 1920.

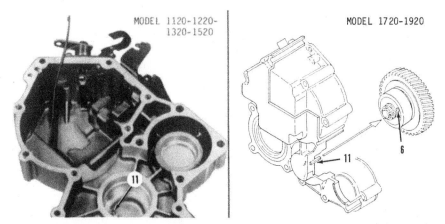

Fig. 64—When installing timing gear case, be sure to align hole in oil pump cover (6) with dowel pin (11) in case.

disconnect inlet and pressure tubes from hydraulic shuttle pump, then unbolt and remove pump from engine. Remove spacer plate (2—Fig. 66), pump drive gear (7) and bearings (6) from timing gear case.

Remove crankshaft pulley using a suitable puller. Unbolt and remove timing gear case from engine.

To reinstall timing gear case, reverse the removal procedure. Make sure that pump drive gear spacer (2—Fig. 66) is installed with grooves (G) positioned at the top. Tighten crankshaft pulley retaining nut to a torque of 274-333 N·m (203-246 ft.-lbs.).

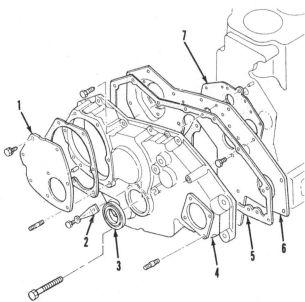

Fig. 65—Exploded view of timing gear cover and related parts used on Model 2120.

1. Hydraulic system pump mounting plate	4. Timing gear case
2. Timing pointer	5. Gasket
3. Crankshaft front oil seal	6. Spacer plate
	7. Gasket

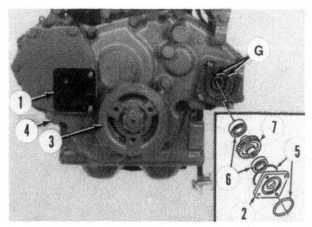

Fig. 66—Front view of Model 2120 timing gear case. When installing spacer plate (4), make certain that grooves (G) in plate are positioned at the top.

1. Hydraulic pump mounting plate	4. Timing gear case
2. Spacer plate	5. "O" rings
3. Crankshaft pulley	6. Bearings
	7. Pump drive gear

TIMING GEARS

Models 1120-1220-1320-1520-1720-1920

35. IDLER GEAR AND OIL PUMP. The rotor type oil pump is located inside the idler timing gear (8—Fig. 67). To remove idler gear and oil pump assembly, first remove timing gear case as outlined in paragraph 33. Remove "E" clip (1) from front of idler gear shaft (4), then slide idler gear with pump components off idler shaft.

Inspect pump cover (6), rotor (7) and port block (10) for wear, scratches or scoring and renew as necessary. Use a feeler gage to check inner rotor to outer rotor clearance as shown in Fig. 68. Standard clear-

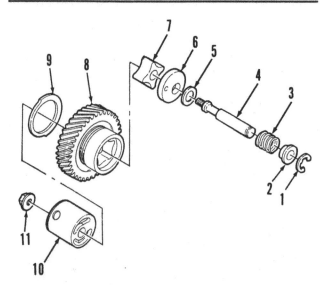

Fig. 67—Engine oil pump is located within the idler timing gear (8) on Models 1120, 1220, 1320, 1520, 1720 and 1920.

1. "E" clip	7. Pump inner rotor
2. Collar	8. Idler gear & pump outer rotor
3. Spring	
4. Idler gear shaft	9. Thrust washer
5. Spacer	10. Oil port block
6. End cover	11. Nut

ance is 0.10-0.15 mm (0.004-0.006 inch). Renew rotors if clearance exceeds 0.25 mm (0.008 inch).

The idler shaft (4—Fig. 69) and oil port block (10) assembly is a press fit in bore of cylinder block. If renewal is required, special tool No. 11097 or other suitable puller tool and a slide hammer may be used to remove oil port block from cylinder block. Special installing tool No. 11063 is available for proper installation of port block. The port block is correctly installed when installing tool bottoms against engine block.

Reinstall thrust washer and idler gear, aligning timing marks on crankshaft gear and camshaft gear with timing marks on idler gear (T—Fig. 70). Assemble oil pump inner rotor, cover, shim, spring, collar and "E" clip on idler shaft. Reinstall timing gear case, making sure that dowel pin in case engages hole in oil pump cover (Fig. 64).

36. CAMSHAFT GEAR AND GOVERNOR. The camshaft gear (6—Fig. 70) is a press fit on camshaft. To remove camshaft gear and governor weight assembly, it is recommended that camshaft be removed from the engine as outlined in paragraph 40. With-

draw governor slider (7) from end of camshaft. Use a suitable press or puller to remove camshaft gear with governor weights from camshaft. Remove key, spacers (3 and 5) and tachometer drive gear (4) from camshaft if necessary.

Inspect camshaft gear teeth and governor weights for excessive wear or damage and renew as necessary.

To reinstall, reverse the removal procedure making sure that timing marks on crankshaft gear, idler gear and camshaft gear are aligned (T—Fig. 70).

Model 2120

37. IDLER GEAR, CAMSHAFT GEAR AND INJECTION PUMP GEAR. To remove timing gears, first remove timing gear case as outlined in paragraph 34. Remove idler gear retaining cap screws and withdraw

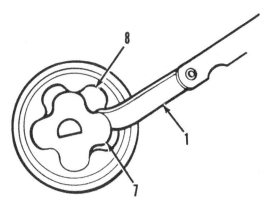

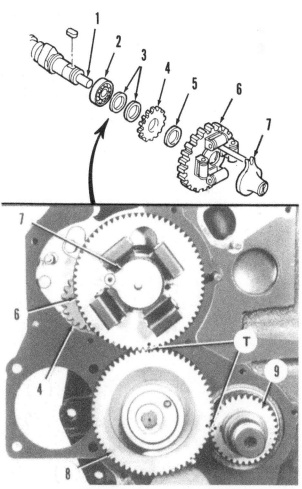

Fig. 68—Use a feeler gage (1) to check clearance between oil pump inner rotor (7) and outer rotor (8).

Fig. 69—Idler gear shaft (4) and oil port block (10) assembly is a press fit in cylinder block.

Fig. 70—View of timing gears used on Models 1120, 1220, 1320, 1520, 1720 and 1920. When installing gears, align timing marks (T) as shown to obtain correct valve timing.

1. Camshaft	6. Camshaft gear &
2. Bearing	governor weight
3. Spacers	assy.
4. Tachometer drive	7. Governor slider
gear	8. Idler gear
5. Spacer	9. Crankshaft gear

plate (4—Fig. 72) and idler gear (3). Remove oil line (9) and idler gear support shaft if necessary. Remove retaining nut (6) from end of camshaft and withdraw camshaft gear (5), spacer washers, tachometer drive gear and collar.

Note alignment marks (A) on injection pump drive gear (1) and drive coupling (2) prior to removing drive gear. If no marks are visible, mark the gear and coupling to facilitate reassembly at original timing setting, then unbolt and remove drive gear.

Inspect timing gears for excessive wear, chipped teeth or other damage and renew as necessary. Check camshaft gear and camshaft keyways and key for wear and renew key, gear and/or camshaft if necessary.

Install idler gear, aligning timing mark on crankshaft gear with mark on idler gear as shown in Fig. 72. Tighten idler gear retaining cap screws to a torque of 24-29 N·m (17-22 ft.-lbs.). Install tachometer drive gear, spacers and camshaft drive gear on the camshaft, aligning timing mark on camshaft gear with mark on idler gear. Tighten camshaft gear retaining nut to a torque of 147-157 N·m (109-116 ft.-lbs.).

Rotate crankshaft counterclockwise (viewed from front) until idler gear-to-crankshaft gear timing mark (T—Fig. 73) is now positioned opposite injection pump drive coupling (2). Install injection pump drive gear (1), aligning timing mark (P) on pump drive gear with timing mark (T) on idler gear. Rotate pump drive coupling as necessary to align timing mark (A) on pump drive coupling with mark on drive gear, then tighten gear retaining cap screws.

NOTE: If new injection pump components or pump drive gear are used, set injection pump timing as outlined in paragraph 57.

CAMSHAFT AND CAM FOLLOWERS

Models 1120-1220-1320-1520-1720-1920

40. REMOVE AND REINSTALL. The camshaft can be removed without separating the engine from transmission. However, if camshaft rear bearing (10—Fig. 74) is to be renewed, engine must be removed from tractor as outlined in paragraph 27 and clutch, flywheel and rear plate removed.

To remove camshaft, first remove radiator, timing gear case and cylinder head. Lift cam followers (12) out from the top of cylinder block. Identify cam followers in order of removal so that they can be reinstalled in their original locations if reused. Remove fuel stop solenoid and fuel injection pump.

Remove governor slider (1—Fig. 75) from end of camshaft. Remove screws from camshaft retaining plate (16), then withdraw camshaft, drive gear and governor weights as an assembly. Use a suitable puller to remove drive gear from camshaft.

Inspect camshaft and cam followers for wear, scoring or other damage. Position camshaft in V-blocks and check runout using a dial indicator at center bearing journal. Maximum allowable runout is 0.10 mm (0.004 inch) for all models. Inspect cam shaft bearings for roughness when rotated, excessive wear or damage and renew as necessary. Refer to the following specifications:

Models 1120-1220
Cam lobe height-valves

Standard	26.445-26.50 mm
	(1.041-1.043 in.)
Minimum	26.1 mm
	(1.027 in.)

Fig. 72—View of Model 2120 timing gear valve train. Align timing marks (T) as shown to correctly time valves.

1. Injection pump drive gear	6. Retaining nut
2. Pump drive hub	7. Crankshaft gear
3. Idler gear	8. Oil pump drive gear
4. Retainer plate	9. Oil line
5. Camshaft gear	

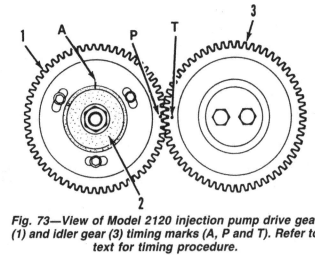

Fig. 73—View of Model 2120 injection pump drive gear (1) and idler gear (3) timing marks (A, P and T). Refer to text for timing procedure.

Cam lobe height-fuel

Standard . . . . .33.94-34.06 mm
(1.335-1.341 in.)
Minimum . . . . .33.8 mm
(1.330 in.)

Models 1320-1520-1720

Cam lobe height-valves

Standard . . . . .34.065-34.120 mm
(1.3411-1.3433 in.)
Minimum . . . . .33.7 mm
(1.327 in.)

Cam lobe height-fuel

Standard . . . . .41.94-42.06 mm
(1.651-1.656 in.)
Minimum . . . . .41.8 mm
(1.646 in.)

Model 1920

Cam lobe height-valves

Standard . . . . .34.48-34.54 mm
(1.3575-1.3598 in.)
Minimum . . . . .34.1 mm
(1.3425 in.)

Cam lobe height-fuel

Standard . . . . .41.94-42.06 mm
(1.651-1.656 in.)
Minimum . . . . .41.8 mm
(1.646 in.)

Fig. 75—An access hole is provided in camshaft gear for removal of retainer plate Allen head bolt (2).

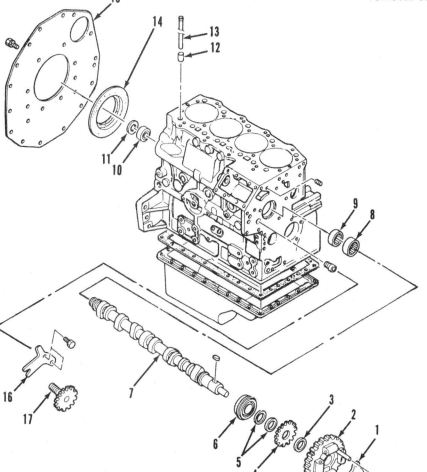

1. Governor slider
2. Camshaft gear & governor weight assy.
3. Spacer
4. Tachometer drive gear
5. Spacers
6. Front bearing
7. Camshaft
8. Front intermediate bearing
9. Rear intermediate bearing
10. Rear bearing
11. Plug
12. Cam follower
13. Push rod
14. Crankshaft rear oil seal
15. Rear plate
16. Retainer plate
17. Tachometer gear

Fig. 74—Exploded view of camshaft and related components used on Model 1920. Models 1120, 1220, 1320, 1520 and 1720 are similar except that only one center bearing is used.

Cam followers should be renewed if camshaft is renewed. Lubricate camshaft, bearings and cam followers prior to installation. Reinstall camshaft and drive gear assembly, aligning timing marks on idler gear with marks on crankshaft gear and camshaft gear as shown in Fig. 70. Install fuel injection pump, and if camshaft was renewed, check and adjust injection pump timing as outlined in paragraph 55.

Model 2120

41. REMOVE AND REINSTALL. To remove camshaft, it is necessary to move the mushroom type cam followers (8—Fig. 76) away from the cam lobes so that camshaft can be withdrawn from front of cylinder block. This is normally accomplished by removing the engine and turning it upside down to allow cam followers to fall away from the camshaft. However, if suitable tools are available to raise and hold the cam followers up, camshaft can be removed without removing engine from the tractor. If camshaft rear bearing (10) is to be renewed, engine must be removed from tractor and clutch, flywheel and rear plate removed from engine.

Drain engine coolant, engine oil, power steering oil and transmission/hydraulic system oil. Remove engine from tractor as outlined in paragraph 27. Remove each of the rocker arm supports and rocker arms. Remove timing gear case as outlined in paragraph 34. Remove idler timing gear, injection pump drive gear, tachometer drive gear and camshaft gear. Unbolt and remove front plate from cylinder block. Turn engine bottom side up and remove the oil pan. Withdraw

camshaft (6) and front bearing (5) from cylinder block. Remove cam followers from bottom side of cylinder block. Identify the position of each cam follower so that they can be reinstalled in original locations if reused. Remove expansion plug (13), spacer (12), snap ring (11) and camshaft rear bearing (10) from rear of cylinder block. Remove needle bearing (9) from center of cylinder block.

Inspect camshaft and cam followers for wear, scoring, pitting or other damage and renew as necessary. Cam followers should be renewed if camshaft is renewed. Position camshaft in V-blocks and check runout using a dial indicator at center bearing journal. Straighten or renew camshaft if runout exceeds 0.10 mm (0.004 inch). Measure camshaft lobe height using a micrometer. Standard height is 38.06-38.07 mm (1.4980-1.499 inches). Renew camshaft if height is less than 37.65 mm (1.482 inches).

To reinstall camshaft, reverse the removal procedure. Lubricate camshaft bearings, journals, cam lobes and cam followers before installation. Align timing gear marks as outlined in paragraph 37.

ROD AND PISTON UNITS

All Models

42. Connecting rod and piston units can be removed from above after removing cylinder head, engine oil pan and oil pump pickup tube. On Models 1920 and 2120 equipped with front wheel drive axle,

1. Nut
2. Camshaft gear
3. Spacer
4. Tachometer drive gear
5. Front bearing
6. Camshaft
7. Push rod
8. Cam follower
9. Center bearing
10. Rear bearing
11. Snap ring
12. Spacer
13. Expansion plug
14. Rear plate
15. Retainer plate
16. Tachometer driven gear

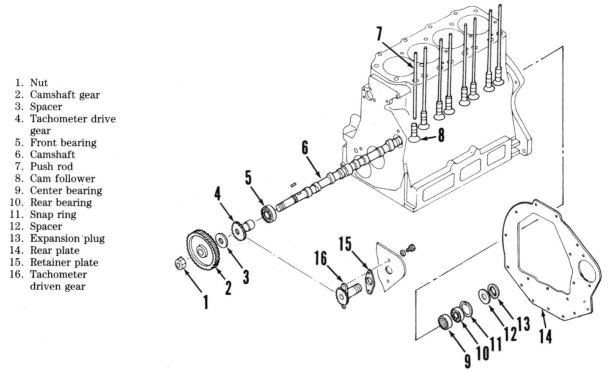

Fig. 76—Exploded view of camshaft and related components used on Model 2120.

the axle and front support assembly must be separated from the engine as outlined in paragraph 27 prior to removing engine oil pan. On all other models equipped with front wheel drive, the front axle drive shaft must be removed prior to removing engine oil pan. On Models 1920 and 2120, engine balancer assembly must also be removed.

> NOTE: Be sure that carbon deposits and ring wear ridge (if present) are removed from top of cylinder before pushing piston out of cylinder block.

Remove connecting rod cap, then push piston and rod assembly out top of cylinder block. Be sure to keep connecting rod caps and bearing inserts with their respective connecting rods for reassembly in original locations.

Piston pin is a transitional fit in piston. Heating piston in hot water will facilitate removal and installation of the pin. Be sure to identify piston and connecting rod for each cylinder so they can be reinstalled in their original locations.

Piston and rod cap must be assembled correctly to connecting rod for proper operation. On all models except 2120, assemble pistons and connecting rods with matching marks (1—Fig. 77) on rod and cap positioned on same side as trade name "SHIBAURA" embossed on inside of piston skirt (2). Be sure that markings on rod and cap are together and that connecting rod and piston are installed in the cylinder block with connecting rod matching marks facing injection pump side of engine. On Model 2120, combustion chamber (1—Fig. 78) is offset slightly in piston crown.

Assemble piston to connecting rod so that combustion chamber is offset toward injection pump side of engine and matching dimples (2) in connecting rod and cap face front of engine. Tighten connecting rod cap retaining bolts to a torque of 24-27 N·m (17-20 ft.-lbs.) on Models 1120 and 1220; 49-54 N·m (36-40 ft.-lbs.) on Models 1320, 1520, 1720 and 1920; 78-83 N·m (58-62 ft.-lbs.) on Model 2120.

PISTONS, RINGS AND CYLINDERS

All Models

43. The aluminum cam ground pistons are fitted with two compression rings and one oil control ring. Pistons and rings are available in standard size and 0.5 mm (0.020 inch) and 1.0 mm (0.040 inch) oversizes.

Clean pistons in suitable solvent and inspect for excessive wear, scratches, scoring or other damage. To check piston ring lands for wear, insert a new ring into piston ring groove and measure side clearance using a feeler gage as shown in Fig. 79. On Model 2120, renew piston if side clearance exceeds 0.30 mm (0.012 inch) for compression rings or 0.15 mm (0.006 inch) for oil control ring. On all other models, renew piston if clearance exceeds 0.25 mm (0.010 inch) for compression rings or 0.15 mm (0.006 inch) for oil control ring.

Check cylinder bore for wear, scoring or other damage. Measure the diameter of cylinder bore in four positions using a suitable cylinder bore gage as shown in Fig. 80. If any of the measurements indicate ex-

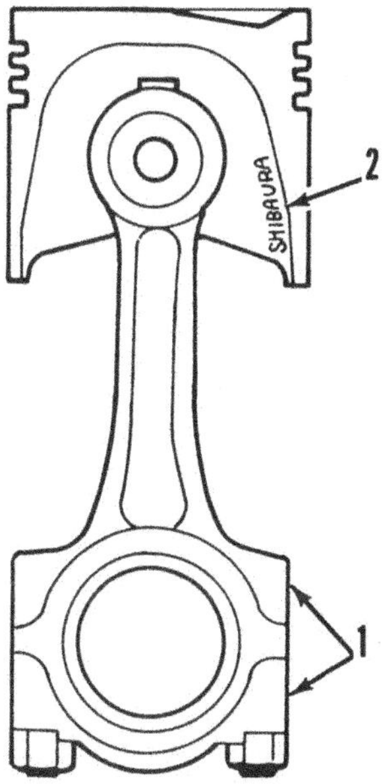

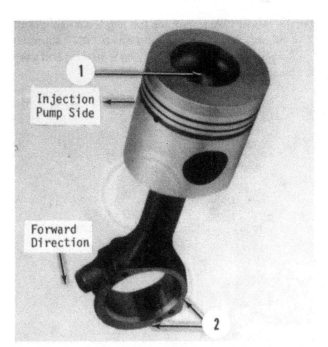

Fig. 77—On all models except Model 2120, assemble piston to connecting rod so that match marks (1) on connecting rod and cap are on same side as "SHIBAURA" trade name (2) embossed on inside of piston skirt.

Fig. 78—On Model 2120, assemble piston to connecting rod so that combustion chamber (1) is offset toward injection pump side of engine and matching dimples in connecting rod and cap face front of engine.

cessively worn or out-of-round cylinder bores, the cylinders should be bored and oversize pistons installed. All cylinders must be bored to the same size.

Check new piston rings in the bore in which they will be installed to ensure that rings have sufficient end gap clearance. Position rings, one at a time, in cylinder bore to lowest point of ring travel using an inverted piston to square the ring in cylinder bore. Use a feeler gage to measure end gap (Fig. 81). Refer to specification table for specified end gap.

Model 1120

Piston skirt diameter
Standard .71.917-71.932 mm
(2.8313-2.8319 in.)
Minimum . . . . .71.7 mm
(2.823 in.)

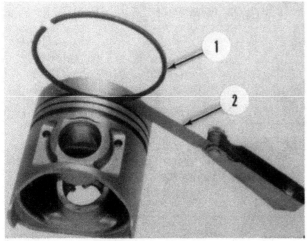

Fig. 79—To check piston ring land wear, position new ring (1) in piston ring groove and measure side clearance with a feeler gage (2).

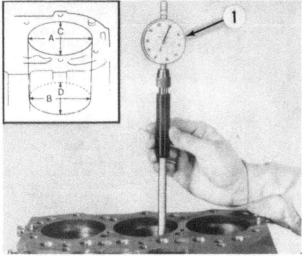

Fig. 80—To check cylinders for taper (wear) and out-of-round, use a suitable bore gage (1) to measure cylinder in four positions (A, B, C and D).

Cylinder bore
Standard .71.990-72.005 mm
(2.8342-2.8348 in.)
Maximum . . . . .72.2 mm
(2.8425 in.)

Piston to cylinder clearance
Standard . . . . .0.0575-0.0875 mm
(0.0022-0.0034 in.)
Maximum . . . . .0.25 mm
(0.010 in.)

Piston pin bore
Standard . . . . .20.998-21.002 mm
(0.8267-0.8269 in.)
Maximum . . . . .21.016 mm
(0.827 in.)

Piston ring-to-groove side clearance
Top ring . . . . .0.06-0.10 mm
(0.0024-0.0039 in.)
Middle ring .0.05-0.09 mm
(0.002-0.004 in.)
Maximum . . . . .0.25 mm
(0.010 in.)
Oil control ring . . . . .0.02-0.06 mm
(0.001-0.0025 in.)
Maximum . . . . .0.15 mm
(0.006 in.)

Piston ring end gap
Top ring . . . . .0.15-0.27 mm
(0.006-0.0105 in.)
Middle ring .0.12-0.24 mm
(0.005-0.0095 in.)
Oil control ring . . . . .0.20-0.35 mm
(0.008-0.014 in.)

Model 1220

Piston skirt diameter
Standard . . . . .74.9425-74.9575 mm
(2.9505-2.9510 in.)
Minimum . . . . .74.7 mm
(2.941 in.)

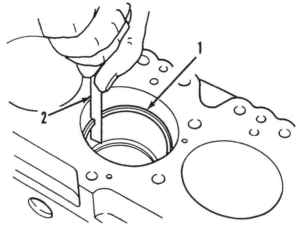

Fig. 81—Insert piston rings (1) squarely into cylinder and measure end gap using a feeler gage (2).

Cylinder bore
    Standard. . . . . .75.0-75.019 mm
        (2.9527-2.9534 in.)
    Maximum . . . . .75.2 mm
        (2.9606 in.)
Piston to cylinder clearance
    Standard . . . . .0.0425-0.0765 mm
        (0.0016-0.0030 in.)
    Maximum . . . . .0.25 mm
        (0.010 in.)
Piston pin bore
    Standard . . . . .20.998-21.002 mm
        (0.8267-0.8269 in.)
    Maximum . . . . .21.016 mm
        (0.827 in.)
Piston ring-to-groove side clearance
    Top ring .0.06-0.10 mm
        (0.0025-0.004 in.)
    Middle ring . . . . .0.05-0.09 mm
        (0.002-0.0035 in.)
    Maximum . . . . .0.25 mm
        (0.010 in.)
    Oil control ring . . . . .0.02-0.06 mm
        (0.001-0.0025 in.)
    Maximum. . . . . .0.15 mm
        (0.006 in.)
Piston ring end gap
    Top ring .0.20-0.35 mm
        (0.008-0.014 in.)
    Middle ring . . . . .0.15-0.30 mm
        (0.006-0.012 in.)
    Oil control ring . . . . .0.15-0.35 mm
        (0.006-0.014 in.)

Models 1320-1520
Piston skirt diameter (1320)
    Standard. . . . . .81.913-81.928 mm
        (3.225-3.226 in.)
    Minimum . . . . .81.7 mm
        (3.216 in.)
Piston skirt diameter (1520)
    Standard . . . . .83.913-83.928 mm
        (3.303-3.304 in.)
    Minimum . . . . .83.7 mm
        (3.295 in.)
Cylinder bore (1320)
    Standard . . . . .82.0-82.019 mm
        (3.228-3.229 in.)
    Maximum . . . . .83.2 mm
        (3.276 in.)
Cylinder bore (1520)
    Standard . . . . .84.0-84.019 mm
        (3.307-3.308 in.)
    Maximum . . . . .85.2 mm
        (3.354 in.)
Piston to cylinder clearance
    Standard . . . . .0.088-0.106 mm
        (0.0034-0.0041 in.)

    Maximum. . . . . .0.30 mm
        (0.012 in.)
Piston pin bore
    Standard . . . . .24.999-25.003 mm
        (0.984-0.9843 in.)
    Maximum . . . . .25.0 mm
        (0.985 in.)
Piston ring-to-groove side clearance
    Top ring .0.07-0.11 mm
        (0.0028-0.0043 in.)
    Middle ring . . . . .0.04-0.08 mm
        (0.0016-0.0031 in.)
    Maximum . . . . .0.25 mm
        (0.010 in.)
    Oil control ring . . . . .0.02-0.06 mm
        (0.001-0.0025 in.)
    Maximum. . . . . .0.15 mm
        (0.006 in.)
Piston ring end gap
    Top ring .0.20-0.35 mm
        (0.008-0.014 in.)
    Middle ring . . . . .0.20-0.40 mm
        (0.008-0.016 in.)
    Oil control ring . . . . .0.20-0.40 mm
        (0.008-0.016 in.)

Model 1720
Piston skirt diameter
    Standard . . . . .83.948-83.963 mm
        (3.3050-3.3056 in.)
    Minimum . . . . .83.7 mm
        (3.295 in.)
Cylinder bore
    Standard . . . . .84.0-84.019 mm
        (3.307-3.308 in.)
    Maximum . . . . .85.2 mm
        (3.354 in.)
Piston to cylinder clearance
    Standard . . . . .0.038-0.064 mm
        (0.0015-0.0025 in.)
    Maximum . . . . .0.25 mm
        (0.010 in.)
Piston pin bore
    Standard . . . . .27.999-28.003 mm
        (1.1023-1.1025 in.)
    Maximum. .28.02 mm
        (1.1031 in.)
Piston ring-to-groove side clearance
    Top ring .0.07-0.11 mm
        (0.0028-0.0043 in.)
    Middle ring . . . . .0.04-0.08 mm
        (0.0016-0.0031 in.)
    Maximum . . . . .0.25 mm
        (0.010 in.)
    Oil control ring . . . . .0.02-0.06 mm
        (0.001-0.0025 in.)
    Maximum . . . . .0.15 mm
        (0.006 in.)

Piston ring end gap
Top ring . . . . .0.20-0.35 mm
(0.008-0.014 in.)
Middle ring . . . . .0.20-0.40 mm
(0.008-0.016 in.)
Oil control ring . . . . .0.20-0.40 mm
(0.008-0.016 in.)

Model 1920
Piston skirt diameter
Standard . . . . .83.943-83.958 mm
(3.3048-3.3054 in.)
Minimum . . . . .83.7 mm
(3.295 in.)
Cylinder bore
Standard . . . . .84.0-84.019 mm
(3.307-3.308 in.)
Maximum . . . . .85.2 mm
(3.354 in.)
Piston to cylinder clearance
Standard . . . . .0.042-0.076 mm
(0.0017-0.003 in.)
Maximum . . . . .0.25 mm
(0.010 in.)
Piston pin bore
Standard . . . . .27.999-28.003 mm
(1.1023-1.1025 in.)
Maximum . . . . .28.02 mm
(1.1031 in.)
Piston ring-to-groove side clearance
Top ring . . . . .0.07-0.11 mm
(0.0028-0.0043 in.)
Middle ring . . . . .0.04-0.08 mm
(0.0016-0.0031 in.)
Maximum . . . . .0.25 mm
(0.010 in.)
Oil control ring . . . . .0.02-0.06 mm
(0.001-0.0025 in.)
Maximum . . . . .0.15 mm
(0.006 in.)
Piston ring end gap
Top ring .0.20-0.35 mm
(0.008-0.014 in.)
Middle ring . . . . .0.20-0.40 mm
(0.008-0.016 in.)
Oil control ring . . . . .0.20-0.40 mm
(0.008-0.016 in.)

Model 2120
Piston skirt diameter
Standard . . . . .84.83-84.91 mm
(3.340-3.343 in.)
Minimum . . . . .84.72 mm
(3.335 in.)
Cylinder bore
Standard . . . . .85.0-85.022 mm
(3.346-3.347 in.)

Maximum . . . . .86.2 mm
(3.394 in.)
Piston to cylinder clearance
Standard . . . . .0.087-0.139 mm
(0.0034-0.0055 in.)
Maximum . . . . .0.30 mm
(0.012 in.)
Piston pin bore
Standard . . . . .32.0 mm
(1.260 in.)
Maximum . . . . .32.01 mm
(1.2603 in.)
Piston ring-to-groove side clearance
Top ring . . . . .0.04-0.08 mm
(0.0016-0.0032 in.)
Middle ring . . . . .0.04-0.08 mm
(0.0016-0.0031 in.)
Maximum . . . . .0.30 mm
(0.012 in.)
Oil control ring . . . . .0.02-0.06 mm
(0.001-0.0025 in.)
Maximum . . . . .0.15 mm
(0.006 in.)
Piston ring end gap
Top ring .0.20-0.40 mm
(0.008-0.016 in.)
Middle ring . . . . .0.20-0.40 mm
(0.008-0.016 in.)
Oil control ring . . . . .0.20-0.40 mm
(0.008-0.016 in.)

Refer to appropriate Fig. 82 or 83 for installation of piston rings. Joint of oil ring expander (1) should be positioned opposite (180 degrees) from oil ring end gap (2). Use a suitable ring expander to install rings onto pistons. Stagger ring end gaps approximately 120 degrees from each other. Do not position a ring gap over piston pin bore. Lubricate cylinder bore, piston and rings with engine oil before installing piston and rod assemblies.

PISTON PIN

All Models

44. The full floating piston pin is retained in piston by snap rings. The pin is a transitional fit in piston (pin diameter may be equal to or slightly larger than piston pin bore at room temperature). Maximum allowable clearance in piston bosses is 0.02 mm (0.0008 inch) for Models 1120, 1220, 1320 and 1520 or 0.04 mm (0.0016 inch) for Models 1720, 1920 and 2120. Renew piston and/or pin if clearance is excessive.

NOTE: Heating piston in hot water will facilitate removal and installation of piston pin.

Piston pin bushing in connecting rod is renewable. After new bushing is pressed into position, an oil hole must be drilled in top of bushing using the hole in

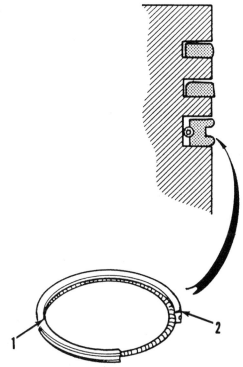

Fig. 82—Cross section of piston and rings showing correct installation of piston rings for Models 1120, 1220, 1320, 1520, 1720 and 1920. Joint of expander spring (1) should be positioned 180 degrees from end gap (2) of oil ring.

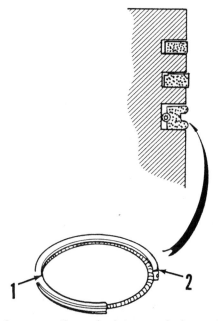

Fig. 83—Cross section of piston and rings showing correct installation of piston rings on Model 2120. Joint of expander spring (1) should be positioned 180 degrees from end gap (2) of oil ring.

upper end of connecting rod as a guide. Bushing must be reamed and honed to provide desired clearance for piston pin.

Refer to the following specification data.

Models 1120-1220
Pin diameter

Standard ....20.998-21.002 mm
(0.8267-0.8268 in.)

Minimum ....20.98 mm
(0.826 in.)

Pin to rod bushing clearance (1120)

Standard ....0.013-0.028 mm
(0.0005-0.0011 in.)

Maximum.........................0.08 mm
(0.003 in.)

Pin to rod bushing clearance (1220)

Standard ....0.008-0.023 mm
(0.0003-0.0009 in.)

Maximum.........................0.08 mm
(0.003 in.)

Models 1320-1520
Pin diameter

Standard ....24.996-25.0 mm
(0.984-0.9843 in.)

Minimum ....24.98 mm
(0.9834 in.)

Pin to rod bushing clearance

Standard ....0.01-0.025 mm
(0.0004-0.001 in.)

Maximum.........................0.08 mm
(0.003 in.)

Models 1720-1920
Pin diameter

Standard ....27.996-28.0 mm
(1.1022-1.1024 in.)

Minimum ....27.98 mm
(1.1012 in.)

Pin to rod bushing clearance

Standard ....0.01-0.025 mm
(0.0004-0.001 in.)

Maximum.........................0.08 mm
(0.003 in.)

Model 2120
Pin diameter

Standard ....31.99-32.0 mm
(1.2595-1.260 in.)

Minimum ....31.97 mm
(1.259 in.)

Pin to rod bushing clearance

Standard ....0.02-0.04 mm
(0.0008-0.0016 in.)

Maximum.........................0.15 mm
(0.006 in.)

CONNECTING RODS AND BEARINGS

All Models

46. Renewable, precision bearing inserts are used in big end of connecting rods. Bearing inserts are available in standard size and 0.25 mm (0.010 inch) and 0.50 mm (0.020 inch) undersizes. A renewable bushing is used in small end of connecting rods. Refer to paragraph 44 for installing and sizing small end bushing.

Check connecting rods for damage and alignment and renew as necessary. A suitable connecting rod alignment fixture should be used to check for bent or twisted connecting rods. Straighten or renew connecting rods as necessary if twist exceeds 0.20 mm (0.008 inch) per 100 mm (3.94 inches) of length, or if bent more than 0.15 mm (0.006 inch) per 100 mm (3.94 inches) of length.

Refer to paragraph 42 for assembly of piston to connecting rod. Check connecting rod bearing clearance using Plastigage. Place a piece of Plastigage (1—Fig. 85) of correct size across the bearing, then install rod cap and tighten to specified torque of 24-27 N·m (17-20 ft.-lbs.) on Models 1120 and 1220; 49-54 N·m (36-40 ft.-lbs.) on Models 1320, 1520, 1720 and 1920; 78-83 N·m (58-62 ft.-lbs.) on Model 2120. Do not rotate crankshaft while Plastigage is positioned on the bearing. Compare width of flattened Plastigage with Plastigage scale (2) to determine bearing clearance. On Models 1120, 1220, 1320, 1520, 1720 and 1920, specified clearance is 0.035-0.083 mm (0.001-0.003 inch) and wear limit is 0.20 mm (0.008 inch). On Model 2120, specified clearance is 0.04-0.10 mm (0.002-0.004 inch) and wear limit is 0.25 mm (0.010 inch).

After installation, push connecting rod to one side and check connecting rod side play (S—Fig. 86) on crankshaft using a feeler gage. Normal side clearance is 0.10-0.30 mm (0.004-0.012 inch) on all models. Renew connecting rod if side clearance exceeds 0.70 mm (0.028 inch).

CRANKSHAFT AND BEARINGS

All Models

47. On Models 1120, 1220, 1320, 1520 and 1720, crankshaft is supported at the front by a sleeve type bearing pressed into the cylinder block and supported at the center and rear by three split type bearing inserts and holders. On Model 1920, crankshaft is supported at the front by a sleeve type bearing pressed into cylinder block and supported at the center and rear by four split type bearing inserts and holders. On Model 2120, crankshaft is supported at front and rear by sleeve type bearings pressed into cylinder block at the front and a bearing retainer plate at rear of cylinder block. Three split type bearing inserts and holders support center main journals.

On Model 2120, crankshaft end play is limited by thrust washers (1—Fig. 90) located at front and rear faces of center main bearing holder. On all other models, crankshaft end play is limited by thrust washers (1—Fig. 88 or 89) located at front and rear faces of rear main bearing holder.

To remove crankshaft and bearings, the engine must be removed from tractor as outlined in paragraph 27. Remove oil pan, engine balancer (Models 1920 and 2120) and piston and connecting rod assemblies. Remove crankshaft pulley and timing gear case. Remove clutch assembly. Loosen but do not remove flywheel retaining bolts. Using a brass drift and hammer, tap end of crankshaft to loosen flywheel from crankshaft flange, then remove bolts and flywheel. On Model 2120, unbolt and remove crankshaft rear bearing and seal retainer from cylinder block. On all other models, remove engine rear plate and crankshaft rear oil seal. Remove cap screws retaining main bearing holders in cylinder block (Fig. 87), then withdraw crankshaft assembly from the rear of engine.

Unbolt and separate main bearing holders with bearing inserts from crankshaft. Identify location of

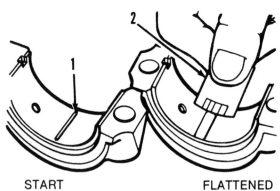

START   FLATTENED
Fig. 85—Connecting rod and main bearing clearance may be checked using Plastigage.

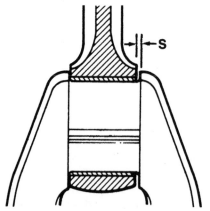

Fig. 86—Check side clearance (S) between connecting rod and crankshaft using a feeler gage.

all components as they are removed so that they can be reinstalled in original positions.

Measure crankshaft main journals and crankpins for size, taper and out-of-round. Crankshaft should be reground to appropriate undersize or renewed if

Fig. 87—View of Model 1920 main bearing holders and retaining bolts (1).

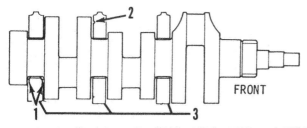

Fig. 88—On Models 1120, 1220, 1320, 1520 and 1720, note that center bearing holder has an identification mark (2) and chamfered side (3) of bearing holders should be toward the front.

taper or out-of-round exceeds 0.05 mm (0.002 inch). Main bearings and connecting rod bearings are available in undersizes of 0.25 mm (0.010 inch) and 0.50 mm (0.020 inch) as well as standard size. Check crankshaft runout and straighten or renew crankshaft if runout exceeds 0.05 mm (0.002 inch).

Refer to the following crankshaft specifications.

Models 1120-1220
Main journal diameter
  Standard . . . . .45.964-45.975 mm
                                                   (1.8096-1.810 in.)
Main bearing clearance
  Standard . . . . .0.039-0.106 mm
                                                   (0.0015-0.004 in.)
  Maximum . . . . .0.20 mm
                                                   (0.008 in.)
Crankpin diameter
  Standard . . . . .38.964-38.975 mm
                                                   (1.534-1.5344 in.)
Crankpin bearing clearance
  Standard . . . . .0.035-0.083 mm
                                                   (0.001-0.003 in.)
  Maximum . . . . .0.20 mm
                                                   (0.008 in.)
Crankshaft end play
  Standard .0.05-0.30 mm
                                                   (0.002-0.011 in.)
  Maximum. . . . . .0.50 mm
                                                   (0.019 in.)
Thrust washer thickness
  Standard . . . . .1.95-2.0 mm
                                                   (0.077-0.079 in.)
  Wear limit . . . . .1.8 mm
                                                   (0.071 in.)

Models 1320-1520
Main journal diameter
  Standard . . . . .57.957-57.970 mm
                                                   (2.281-2.282 in.)

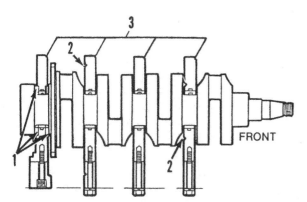

Fig. 89—On Model 1920, front bearing holder and second (from rear) bearing holder have an identification mark (2) and chamfered side (3) of bearing holders should be toward the front.

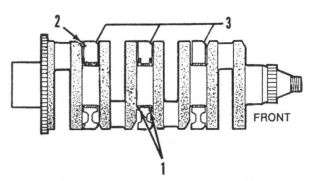

Fig. 90—On Model 2120, rear bearing holder has an identification mark (2) and chamfered side (3) of bearing holders should be toward the front.

Main bearing clearance
   Standard . . . . .0.044-0.102 mm
                  (0.0017-0.0040 in.)
   Maximum .0.20 mm
                  (0.008 in.)

Crankpin diameter
   Standard . . . . .43.964-43.975 mm
                  (1.730-1.731 in.)

Crankpin bearing clearance
   Standard . . . . .0.035-0.083 mm
                  (0.001-0.003 in.)
   Maximum .0.20 mm
                  (0.008 in.)

Crankshaft end play
   Standard . . . . .0.10-0.40 mm
                  (0.004-0.016 in.)
   Maximum .0.50 mm
                  (0.019 in.)

Thrust washer thickness
   Standard . . . . .2.95-3.0 mm
                  (0.116-0.118 in.)
   Wear limit . . . . .2.8 mm
                  (0.110 in.)

Models 1720-1920
Main journal diameter
   Standard. . . . . .67.96-67.97 mm
                  (2.6755-2.676 in.)

Main bearing clearance
   Standard . . . . .0.044-0.102 mm
                  (0.0017-0.0040 in.)
   Maximum .0.20 mm
                  (0.008 in.)

Crankpin diameter
   Standard . . . . .51.964-51.975 mm
                  (2.0458-2.0463 in.)

Crankpin bearing clearance
   Standard . . . . .0.035-0.083 mm
                  (0.001-0.003 in.)
   Maximum .0.20 mm
                  (0.008 in.)

Crankshaft end play
   Standard . . . . .0.10-0.40 mm
                  (0.004-0.016 in.)
   Maximum. . . . . .0.50 mm
                  (0.019 in.)

Thrust washer thickness
   Standard . . . . .2.95-3.0 mm
                  (0.116-0.118 in.)
   Wear limit . . . . .2.8 mm
                  (0.110 in.)

Model 2120
Main journal diameter
   Standard. . . . . .67.96-67.97 mm
                  (2.6755-2.676 in.)

Main bearing clearance
   Standard . . . . .0.056-0.131 mm
                  (0.002-0.005 in.)
   Maximum .0.20 mm
                  (0.008 in.)

Crankpin diameter
   Standard . . . . .59.95-59.97 mm
                  (2.360-2.361 in.)

Crankpin bearing clearance
   Standard . . . . .0.04-0.10 mm
                  (0.002-0.004 in.)
   Maximum .0.20 mm
                  (0.008 in.)

Crankshaft end play
   Standard . . . . .0.10-0.45 mm
                  (0.004-0.018 in.)
   Maximum .0.70 mm
                  (0.028 in.)

Thrust washer thickness
   Standard . . . . .2.435-2.5 mm
                  (0.095-0.098 in.)
   Wear limit . . . . .2.3 mm
                  (0.091 in.)

Measure inside diameter of sleeve type bearing in front of cylinder block (all models) and rear sleeve bearing in bearing retainer (Model 2120). Renew sleeve bearings if inside diameter exceeds 46.10 mm (1.815 inches) on Models 1120 and 1220; 58.15 mm (2.289 inches) on Models 1320 and 1520; 68.15 mm (2.683 inches) on Models 1720, 1920 and 2120.

Check main bearing clearance using Plastigage (Fig. 85). Position piece of Plastigage of correct size across main bearing insert, then install insert and bearing holder assembly on crankshaft. Tighten bearing holder cap screws to specified torque of 25-29 N·m (18-22 ft.-lbs.) on Models 1120 and 1220; 49-54 N·m (36-40 ft.-lbs.) on Models 1320, 1520, 1720 and 1920; 71-81 N·m (51-58 ft.-lbs.) on Model 2120. Do not allow bearing holder to rotate while tightening retaining cap screws as Plastigage will be damaged. Remove bearing holder and measure width of flattened Plastigage using Plastigage scale to determine bearing clearance.

Lubricate bearing inserts and crankshaft, then assemble bearings, thrust washers and bearing holders on the crankshaft. Be sure that grooved side of thrust washers faces crankshaft mating surface. Chamfered side of bearing holders must face toward front of crankshaft. On Models 1120, 1220, 1320, 1520 and 1720 second from rear bearing holder has an identification mark (2—Fig. 88) on it to ensure correct reassembly. On Model 1920, second bearing holder from rear and front bearing holder have identification marks (2—Fig. 89). On Model 2120, rear bearing holder has an identifying mark (2—Fig. 90). Tighten bearing holder cap screws to specified torque as outlined above, then install crankshaft assembly in cylinder block from the rear.

Install bearing holder retaining cap screws and tighten to specified torque of 25-29 N·m (18-22 ft.-lbs.) on Models 1120 and 1220; 49-54 N·m (36-40 ft.-lbs.) on Models 1320, 1520, 1720 and 1920; 78-81 N·m (51-58 ft.-lbs.) on Model 2120. Use a dial indicator to check crankshaft end play. If end play exceeds specified limits, renew thrust washers.

On all models except 2120, install crankshaft rear seal and engine rear plate. Tighten rear plate cap screws to a torque of 27-33 N·m (20-25 ft.-lbs.) on Models 1120 and 1220; 13-17 N·m (10-13 ft.-lbs.) on Models 1320, 1520, 1720 and 1920. On Model 2120, install rear bearing and seal retainer housing and tighten retaining cap screws to a torque of 46-54 N·m (34-40 ft.-lbs.). On Models 1920 and 2120, install and time engine balancer assembly as outlined in paragraph 48. On all models, complete installation by reversing the removal procedure. Be sure that timing gear marks are correctly aligned as outlined in paragraph 35 or 37.

ENGINE BALANCER

Models 1920-2120

48. R&R AND OVERHAUL. To remove engine balancer assembly, first drain oil from engine. If equipped with front wheel drive, remove front axle and axle rear support from tractor as outlined in paragraph 27. Remove engine oil pan and oil pickup tube and screen. Remove balancer oil supply tube. Remove balancer retaining cap screws and lower balancer from cylinder block. Note location and thickness of shims (2—Fig. 91) so that they can be reinstalled in original position.

Prior to disassembling balancer, check balancer gears for timing marks. If no marks are visible, scribe reference marks on each gear so that they can be correctly timed when reassembled. Check backlash between balancer gears. Renew balancer gears and shafts if backlash between gears exceeds 0.15 mm (0.006 inch).

To disassemble, unbolt and remove retaining plates (5—Fig. 91). Loosen set screws in counterweights (7), then withdraw drive gear and shaft assemblies (1) from counterweights and balancer housing. Inspect bushings (6) and shafts for excessive wear or damage and renew as necessary. Make certain that oil hole in bushings is aligned with oil passage in balancer housing. If bushings have rotated in housing bores, renew balancer housing. Lubricate bushings with white lithium grease prior to reassembly.

To reassemble balancer unit, reverse the disassembly procedure. Be sure to align timing marks on balancer gears. Tighten balancer counterweight set

screws to specified torque of 49-54 N·m (36-40 ft.-lbs.) on Model 1920 or 24-29 N·m (17-22 ft.-lbs.) on Model 2120.

To reinstall balancer, rotate crankshaft to position number one and four pistons at top dead center. Position balancer counterweights straight down, then install balancer unit with original shims (2). Tighten balancer retaining cap screws to a torque of 49-54 N·m (36-40 ft.-lbs.) on Model 1920 or 75-81 N·m (55-60 ft.-lbs.) on Model 2120. Measure backlash between balancer drive gear and crankshaft gear. Backlash should not exceed 0.05 mm (0.002 inch) on Model 1920 or 0.06 mm (0.0024 inch) on Model 2120. If necessary, remove or install shims (2) to obtain desired backlash.

FLYWHEEL

All Models

49. To remove flywheel, first split tractor between engine and clutch housing as outlined in paragraph 87. Remove clutch assembly, and loosen but do not remove flywheel retaining cap screws. Using a brass drift and hammer, tap on end of crankshaft to loosen flywheel from crankshaft flange, then remove cap screws and flywheel.

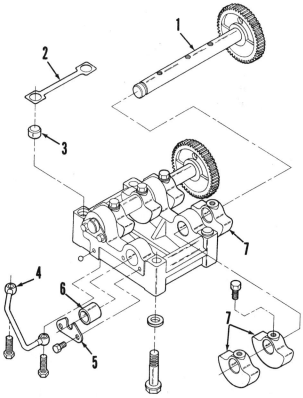

Fig. 91—Exploded view of engine balancer assembly used on Models 1920 and 2120.

1. Balancer shaft & gear	4. Oil tube
2. Shim	5. Retainer plate
3. Dowel pin	6. Bushing
	7. Counterweights

Inspect flywheel and ring gear for excessive wear or other damage and renew if necessary. To remove starter ring gear, use a hammer and chisel to crack the ring gear between two teeth. Heat new ring gear to a temperature of 120°-150° C (250°-300° F). Install heated gear onto flywheel making sure that it is seated against shoulder of flywheel, then quench gear quickly to obtain a shrink fit on the flywheel.

To reinstall flywheel, reverse the removal procedure. Tighten retaining cap screws to specified torque of 59-69 N·m (44-51 ft.-lbs.) on all models.

OIL PUMP AND RELIEF VALVE

Models 1120-1220-1320-1520-1720-1920

50. The gerotor type oil pump is located inside the engine idler gear. Refer to paragraph 35 for service procedures covering the oil pump.

Engine oil pressure relief valve is located in right side of cylinder block directly below the fuel injection pump. Relief valve is serviced as an assembly and is not adjustable. To check oil pressure, install a pressure gage in place of oil pressure sender unit. Relief valve opening pressure should be 290-455 kPa (42-66 psi) at rated engine speed.

Model 2120

51. The gear type oil pump is located in the front of the engine block (Fig. 92). To remove oil pump cover (4) and gears (5 and 6), first remove timing gear case as outlined in paragraph 34. Remove pump drive gear retaining nut and pull drive gear off pump shaft. Unbolt and remove pump cover (4) and withdraw gear set from pump body.

To remove pump body (7), remove engine oil pan. Remove oil pickup tube (9) and screen (10), then withdraw pump body from front of engine block.

Inspect pump gears, body and cover for wear tracks, scratches or scoring and renew as necessary. Use a straightedge and feeler gage to measure gear to housing end clearance as shown in Fig. 93. Specified clearance is 0.05-0.12 mm (0.002-0.0047 inch). Renew pump assembly if clearance exceeds 0.15 mm (0.006 inch). Use a feeler gage to measure gear tip to housing clearance as shown in Fig. 94. Renew pump assembly if clearance exceeds 0.15 mm (0.006 inch).

To reinstall oil pump, reverse the removal procedure. Tighten oil pump drive gear retaining nut to specified torque of 29-34 N·m (22-25 ft.-lbs.). Tighten oil pan retaining screws to torque of 14-19 N·m (10-14 ft.-lbs.).

Oil pressure relief valve (2—Fig. 92) is located in the front of engine block next to the oil pump. Relief valve is serviced as an assembly and is not adjustable. Relief valve opening pressure can be checked by installing a pressure gage in place of oil pressure sender unit. Relief pressure should be 241-393 kPa (35-57 psi) at rated engine speed. Minimum oil pressure at low idle speed with oil temperature of 80° C (175° F) is 117 kPa (17 psi).

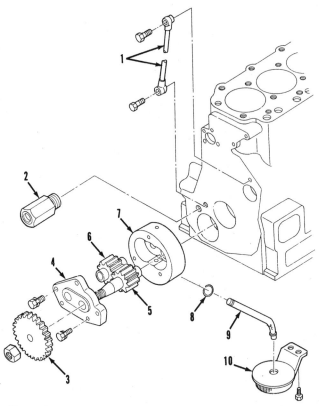

Fig. 92—Exploded view of gear type engine oil pump used on Model 2120.

1. Oil pressure line
2. Pressure relief valve
3. Drive gear
4. Pump cover
5. Drive gear & shaft
6. Idler gear
7. Pump body
8. "O" ring
9. Oil pick-up tube
10. Screen

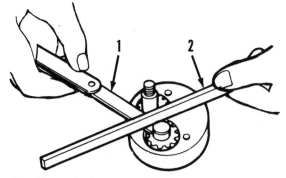

Fig. 93—Use a feeler gage (1) and straightedge (2) to check end clearance of oil pump gears.

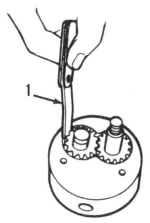

Fig. 94—Measure clearance between gear teeth and body using a feeler gage as shown.

DIESEL FUEL SYSTEM

All models are equipped with in-line type fuel injection pumps having one pumping element for each cylinder. On Model 2120, the pumping elements are actuated by a camshaft enclosed in the pump housing. On all other models, the pumping elements are actuated by cam lobes on the engine valve train camshaft.

Because of extremely close tolerances and precise requirements of all diesel components, it is of the utmost importance that only clean fuel is used and careful maintenance be practiced at all times. Unless necessary special tools are available, service on injectors and injection pumps should be limited to removal, installation and exchange of complete assemblies.

FUEL FILTER AND BLEEDING

All Models

53. Fuel filter life depends upon careful maintenance as well as the hours of operation. The necessity for careful filling with clean fuel cannot be overstressed.

Fuel filter sediment bowl (6—Fig. 95 or 96) should be drained whenever water or sediment is visible in bowl. Fuel filter element should be renewed after every 200 hours of operation, or sooner if loss of engine power is evident. When installing a new filter, open fuel shut-off valve and allow sediment bowl to fill with fuel prior to tightening sediment bowl retaining ring nut. Bleed air from system as follows:

The fuel system should be bled if fuel tank is allowed to run dry or if fuel lines, filter or other components within system have been removed. To bleed air from system, open fuel shut-off valve and loosen bleed screw (2—Fig. 95 or 96) located on fuel injec-

tion pump. Tighten bleed screw when air-free fuel flows from bleed screw.

If engine fails to start after completing the above bleeding procedure, loosen high pressure fuel lines at the injectors. Move throttle to high speed position, then crank engine with starter until fuel flows from loosened connections. Tighten fuel line connections and start engine.

INJECTION PUMP

The maintenance of absolute cleanliness is of the utmost importance when servicing the injection

Fig. 95—View of fuel filter and air bleed screw typical of all tractors except Model 2120.

1. Fuel inlet line	5. Fuel stop solenoid
2. Bleed screw	6. Fuel filter
3. Injection pump	7. Fuel shut-off
4. Throttle cable	valve

Fig. 96—View of fuel filter and air bleed screw on Model 2120.

2. Bleed screw
6. Fuel filter
7. Fuel shut-off valve

pump. Service work or disassembly of injection pump other than that specified should not be attempted without necessary equipment and training.

Models 1120-1220-1320-1520-1720-1920

54. REMOVE AND REINSTALL. To remove fuel injection pump, first shut off fuel supply. Disconnect fuel inlet line (1—Fig. 95) from pump. Disconnect injector lines and cap all openings. Disconnect throttle control cables (4). Remove fuel stop solenoid (5). Remove pump retaining nuts, then raise injection pump (3) and disconnect governor link from pump control rack. Lift pump from engine block, being careful not to damage or lose shims located between pump mounting flange and engine block. The shims are used to adjust injection pump static timing.

To reinstall pump, reverse the removal procedure. If pump or drive components are being renewed or if timing shim thickness is not known, install a shim of 0.5 mm (0.012 inch) thickness. Refer to paragraph 55 for timing of pump and selection of proper thickness shims. If delivery valve holders loosened during removal of injector lines, tighten holders to specified torque of 39-44 N·m (29-32 ft.-lbs.). Bleed air from system as outlined in paragraph 53.

55. PUMP TIMING. To check and adjust injection pump timing, shut off fuel supply to pump. Remove number one delivery valve holder (1—Fig. 98), spring (2) and delivery valve (4) from injection pump. Rein-

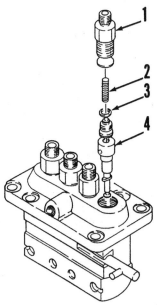

Fig. 98—Exploded view of injection pump delivery valve components typical of all tractors except Model 2120.

1. Delivery valve holder
2. Spring
3. Gasket
4. Delivery valve assy.

stall delivery valve spring and holder and tighten holder finger tight. Turn crankshaft until number one (front) piston is at top dead center on compression stroke, then turn crankshaft counterclockwise approximately 45 degrees. Turn on fuel supply to pump and note that fuel should be flowing from number one delivery valve holder. Slowly rotate crankshaft clockwise to locate exact point at which fuel stops flowing from delivery valve holder, which is beginning of injection. At this point, timing mark on crankshaft pulley should be aligned with the following timing marks on timing gear case: Models 1120 and 1220—23 to 24 degrees BTDC; Models 1320 and 1520—20 to 21 degrees BTDC; Model 1720—22 to 23 degrees BTDC; Model 1920—18 to 19 degrees BTDC.

Injection timing is adjusted by increasing or decreasing thickness of shims located between pump flange and engine block. To change injection timing by 1 degree requires changing shim thickness by approximately 0.10 mm (0.004 inch). Increasing shim thickness will retard timing and decreasing shim thickness will advance timing.

The following pump spill timing procedure may also be used if there is any doubt about accuracy of crankshaft pulley timing marks. Shut off fuel supply to pump, then remove number one delivery valve holder (1—Fig. 98), spring (2) and delivery valve (4) from injection pump. Reinstall delivery valve spring and holder and tighten holder finger tight. Remove rocker arm cover and rotate crankshaft to position number one piston at TDC on compression stroke (both intake and exhaust valve rocker arms loose).

Remove rocker arm shaft support assembly. Remove spring from one of the valves of number one cylinder and allow valve to rest on top of the piston. Position a dial indicator on the valve stem as shown in Fig. 99, turn crankshaft to bring number one piston to exact TDC, then set indicator gage to zero. Turn crankshaft counterclockwise until dial indicator reads 6.35 mm (0.250 inch). Turn on fuel supply and note that fuel should be flowing from number one delivery valve holder. Slowly rotate crankshaft clockwise until fuel just stops flowing. At this point, dial indicator reading should be as follows: Models 1120 and 1220 should be between 3.75-4.07 mm (0.148-0.160 inch) which corresponds to 23-24 degrees BTDC injection timing. Models 1320 and 1520 should be between 3.15-3.46 mm (0.124-0.136 inch) which corresponds to 20-21 degrees BTDC injection timing. Model 1720 should be between 4.27-4.65 mm (0.168-0.183 inch) which corresponds to 22-23 degrees BTDC injection timing. Model 1920 should be between 2.88-3.20 mm (0.113-0.126 inch) which corresponds to 18-19 degrees BTDC injection timing. If timing is not correct, adjust shim thickness as required to obtain specified timing.

Reinstall delivery valve and tighten delivery valve holder to a torque of 39-44 N·m (29-32 ft.-lbs.).

Model 2120

56. REMOVE AND REINSTALL. To remove fuel injection pump, first drain engine coolant and remove grille, radiator side panels and radiator. Scribe alignment marks on pump mounting flange and engine front plate to facilitate correct alignment when reinstalling pump.

Shut off fuel supply and disconnect fuel inlet line (2—Fig. 100) from pump. Remove pump lube oil line (4). Remove injector lines and plug all openings. Dis-

Fig. 99—Injection pump timing may be checked by lowering a valve onto number one piston and measuring piston height using a dial indicator. Refer to text.

1. Dial indicator
2. Valve stem
3. Crankshaft pulley timing mark
4. Timing degrees

Fig. 100—View of injection pump used on Model 2120. Scribe alignment marks (M) on pump flange and engine front plate prior to removing injection pump.

1. Throttle cables
2. Fuel inlet line
3. Fuel stop lever
4. Lube oil line
5. Fuel stop solenoid
6. Turnbuckle

connect throttle cables (1) from pump lever and fuel shut-off cables from fuel stop lever (3). Disconnect inlet and pressure lines from hydraulic system pump. Remove hydraulic pump retaining nuts and withdraw pump from front of timing gear case. Remove injection pump drive gear cover plate from front of timing gear case. Turn crankshaft until timing marks (6—Fig. 101) on idler gear and pump gear are aligned. Scribe alignment marks (4) on pump drive gear (5) and coupling (1). Remove retaining nut (2) and withdraw pump drive gear and coupling as an assembly, being careful not to drop the nut, lockwasher or drive shaft key into timing gear case during removal. Remove injection pump retaining nuts and remove pump from engine.

If injection pump, drive coupling or timing gears are renewed, time pump to engine as outlined in paragraph 57. Install injection pump, aligning scribe marks (M—Fig. 100) on pump flange and engine rear plate made prior to removal. Install pump drive coupler and key and tighten retaining nut to a torque of 39-44 N·m (29-33 ft.-lbs.). With number one piston on compression stroke, rotate crankshaft so that timing mark on idler gear (7—Fig. 101) will align with mark on pump gear (5). Turn pump drive shaft to position notch (4) on drive coupling adjacent to longest slot in pump gear. Install the three gear retaining cap screws and tighten securely. Bleed air from fuel sys-

Fig. 101—View of Model 2120 injection pump drive gear and idler gear timing marks.

1. Pump drive coupling
2. Nut
4. Coupling notch

5. Pump drive gear
6. Timing marks
7. Idler gear
8. Cap screws

tem as outlined in paragraph 53. Check pump to engine timing as outlined in paragraph 57, then install hydraulic pump and radiator.

57. PUMP TIMING. The injection pump timing must be checked if the injection pump or any of the drive components are renewed.

To check pump timing, shut off fuel supply and remove number one delivery valve holder (1—Fig. 102), spring (2) and delivery valve piston (3). Reinstall the spring and delivery valve holder and tighten holder finger tight. Turn crankshaft clockwise until number one piston is at top dead center on compression stroke, then turn crankshaft counterclockwise approximately 45 degrees. Turn on fuel supply to pump and note that fuel should be flowing from number one delivery valve holder. Slowly rotate crankshaft clockwise to locate exact point at which fuel stops flowing from delivery valve holder, which is beginning of injection. At this point, first timing mark (3—Fig. 103) on crankshaft pulley should be aligned with timing pointer (1) on timing gear cover.

The following pump spill timing procedure may also be used if there is any doubt about accuracy of crankshaft pulley timing marks. Remove valve rocker cover. Rotate crankshaft clockwise until number one piston is on compression stroke (intake and exhaust rocker arms loose) and TDC mark (2—Fig. 103) on crankshaft pulley is aligned with timing pointer (1).

Remove spring from one of the valves of number one cylinder and allow valve to rest on top of the piston. Position a dial indicator on valve stem, rotate crankshaft to bring piston to exact TDC, then zero the dial gage as shown in Fig. 104. Rotate crankshaft counterclockwise until dial indicator reads approximately 5.0 mm (0.20 inch). Turn on fuel supply and note that fuel should flow from number one delivery valve holder. Slowly turn crankshaft clockwise to locate exact point at which fuel stops flowing from delivery valve holder and observe dial indicator reading. Pump timing is correct if dial gage reading is within specified range of 3.67-4.05 mm (0.144-0.159 inch) which corresponds to 19.5-20.5 degrees BTDC injection timing.

If pump timing is incorrect, drain engine coolant and remove battery and support bracket, grille, side

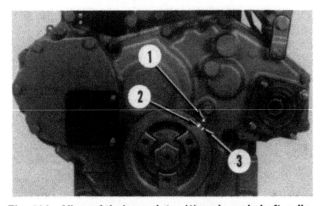

Fig. 103—View of timing pointer (1) and crankshaft pulley top dead center timing mark (2) and injection timing mark (3) on Model 2120.

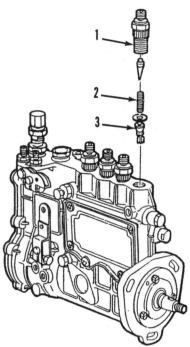

Fig. 102—Exploded view of Model 2120 injection pump delivery valve assembly.

1. Delivery valve holder
2. Spring
3. Delivery valve

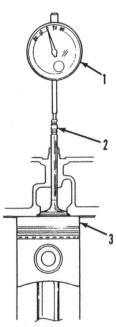

Fig. 104—To check injection pump timing on Model 2120, lower a valve (2) onto top of number 1 piston (3) and use a dial indicator (1) to measure piston height. Refer to text.

panels and radiator. Disconnect inlet and pressure tubes from hydraulic pump, then unbolt and remove pump from timing gear case. Remove injection pump drive gear cover plate from front of timing gear case. Loosen the three pump drive gear retaining cap screws. Rotate crankshaft until beginning of injection mark (3—Fig. 103) on crankshaft pulley is aligned with timing pointer or until dial indicator reading is within specified range of 3.67-4.05 mm (0.144-0.159 inch). While holding pump drive gear from turning, rotate pump drive coupling (1—Fig. 101) to locate exact point at which fuel stops flowing from delivery valve holder. Tighten drive gear retaining screws at the point of fuel cut-off. Use a chisel to mark pump drive coupling and gear alignment for future reference.

Reinstall delivery valve and tighten delivery valve holder to a torque of 30-35 N·m (21-25 ft.-lbs.).

ENGINE SPEED ADJUSTMENT

All Models

58. To adjust engine speed, first start engine and operate until warm. Disconnect throttle cables (1—Fig. 100 or 105) from pump lever.

> NOTE: Be sure that groove in end of governor shaft is aligned with slot in governor lever (4).

Engine speed should be 900-1000 rpm on Models 1920 and 2120 and 800-900 rpm on all other models. If low idle speed is incorrect, loosen locknut and adjust idle speed stop screw (L) as necessary. Move governor lever to maximum speed position. Maxi-

Fig. 105—View of engine speed adjustment points typical of all tractors except Model 2120. Be sure that slot of governor lever is aligned with groove in governor shaft (4).

H. High idle stop
    screw
L. Low idle stop
    screw
1. Throttle cables
6. Turnbuckle

mum no-load speed should be 2650-2700 rpm on all models. To adjust high idle speed, loosen locknut and turn stop screw (H) as necessary. After adjusting high idle stop screw, install a new lock wire on the stop screw.

Reconnect throttle cables and check for correct engine speeds. If necessary, turn turnbuckles (6) to lengthen or shorten cables to obtain full travel of governor lever when using throttle hand control lever or foot control pedal.

GOVERNOR

Models 1120-1220-1320-1520-1720-1920

59. The governor flyweights are located on the forward end of engine camshaft. The governor control linkage is located inside the timing gear case. To service governor linkage or flyweights, refer to paragraph 33 for removal of timing gear case. The flyweights and camshaft gear are serviced as an assembly. Refer to Fig. 63 for exploded view of governor linkage.

When reassembling linkage, align groove in governor arm shaft with slot of governor lever as shown in Fig. 105.

Model 2120

60. The governor assembly is located on rear end of the injection pump camshaft within the injection pump housing. The injection pump and governor should be serviced only by an authorized diesel injection service center.

INJECTOR NOZZLES

All Models

61. LOCATING A FAULTY NOZZLE. If rough or uneven engine operation, or misfiring indicates a faulty injector nozzle, the defective unit can usually be located as follows:

With engine running at the speed where malfunction is most noticeable, loosen compression nut on high pressure line for each injector in turn, allowing fuel to escape at the nut rather than enter the cylinder. If engine operation is not affected when injector line is loosened, that is the cylinder that is misfiring.

If a faulty injector is found and considerable time has elapsed since injectors have been serviced, it is recommended that all injectors be removed and serviced, or that new or reconditioned units be installed. Refer to the following paragraphs 62 through 66 for removal and test procedures.

62. REMOVE AND REINSTALL. Before removing injector lines and injectors, thoroughly clean injectors and surrounding area. Remove fuel leak-off lines and high pressure lines. Cap all openings to prevent entry of dirt. On Models 1120, 1220, 1320, 1520 and 1720, unscrew injector assemblies from cylinder head. On Models 1920 and 2120, remove retaining cap screw or stud nuts from injector clamps and withdraw injector assembly from cylinder head. Be sure to remove injector seal washer from cylinder head bore if it does not come out with the injector.

Precombustion chambers (4—Fig. 107) are fitted in cylinder head on Models 1120 and 1220. The glow plugs must be removed before attempting to remove the prechambers. Use a suitable puller and slide hammer to remove prechambers, or remove cylinder head and tap out prechambers.

If precombustion chambers were removed on Models 1120 and 1220, install new gasket (5), prechamber (4) and heat shield (3) in cylinder head. On all models, install injectors using new seal washers. On Models 1120 and 1220, tighten injectors to specified torque of 79-83 N·m (58-61 ft.-lbs.). On Models 1320, 1520 and 1720, tighten injectors to specified torque of 59-69 N·m (43-51 ft.-lbs.).

Connect fuel lines to injectors, but do not tighten high pressure fuel line fittings at the injectors at this time. Move throttle control to full speed position and crank engine with starter until fuel is discharged at loosened connections, then tighten high pressure line fittings.

63. TESTING. A complete job of testing and adjusting injector nozzles requires use of special test equipment. Nozzle should be tested for opening pressure, seat leakage and spray pattern.

WARNING: Fuel leaves injector nozzle with sufficient force to penetrate the skin. Keep exposed portions of your body clear of nozzle spray when testing.

Before conducting tests, operate tester lever until fuel flows, then attach injector to tester. Operate tester lever a few quick strokes to purge air from injector and to make sure that nozzle valve is not stuck.

64. OPENING PRESSURE. Operate tester lever slowly while observing tester gage reading. Opening pressure should be 11760 kPa (1705 psi) for Models 1120, 1220, 1320 and 1520; 14825 kPa (2150 psi) for Model 1720; 20590 kPa (2985 psi) for Model 1920; 19610 kPa (2845 psi) for Model 2120. On all models, opening pressure is adjusted by adding or removing shims (5—Fig. 108 or 2—Fig. 109).

65. SPRAY PATTERN. Operate tester lever several quick strokes and observe spray pattern. The injector nozzle used on Models 1920 and 2120 has four orifices in nozzle tip. Fuel must be sprayed equally

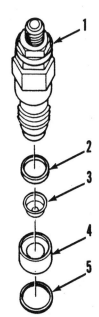

Fig. 107—A precombustion chamber (4) is used on Models 1120 and 1220.

1. Injector assy.
2. Seal washer
3. Heat shield
4. Precombustion chamber
5. Seal washer

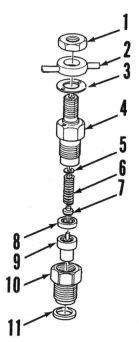

Fig. 108—Exploded view of fuel injector used on Models 1120, 1220, 1320, 1520 and 1720.

1. Nut
2. Fuel return fitting
3. Gasket
4. Nozzle holder
5. Shim
6. Spring
7. Spring seat
8. Spacer
9. Nozzle valve assy.
10. Nozzle nut
11. Seal washer

through the four orifices and be finely atomized and equally spaced. The injector nozzle used in all other models has a single orifice in nozzle tip. Fuel must emerge in a straight line from the orifice and be well atomized and slightly conical. On all models, if pattern is wet, ragged or intermittent, nozzle must be overhauled or renewed.

66. SEAT LEAKAGE. Wipe nozzle tip dry, then operate tester lever to bring gage pressure to 3450 kPa (500 psi) below opening pressure and hold this pressure for 10 seconds. If a drop of fuel forms on nozzle tip, leakage at nozzle valve seat is indicated and nozzle should be overhauled or renewed. Wetting of nozzle tip without formation of a drop is permissible.

67. OVERHAUL. Hard or sharp tools, emery cloth, grinding compound or other than approved solvents must never be used when servicing injectors. An approved nozzle cleaning kit is available through a number of specialized sources.

To disassemble injector used on Models 1120, 1220, 1320, 1520 and 1720, secure nozzle holder (4—Fig. 108) in a soft jawed vise or holding fixture and remove nozzle nut (10). Separate nozzle valve (9), spacer (8), spring seat (7), spring (6) and shims (5) from nozzle holder.

To disassemble injector used on Models 1920 and 2120, secure nozzle holder (1—Fig. 109) in a soft jawed vise or holding fixture and remove nozzle nut (8). If nozzle nut is difficult to remove, soak injector in a carbon removing cleaning solvent for several hours before removing nozzle nut; otherwise nozzle valve may turn with the nut, damaging locating dowel pins (6). Remove nozzle valve (7), spacer (5), spring seat (4), spring (3) and shims (2).

Place all parts in clean calibrating oil or diesel fuel as they are removed. Use a compartmented pan to keep parts from each injector separate from other units if more than one injector is being serviced.

Clean exterior surfaces with a brass wire brush, soaking in an approved carbon solvent if necessary, to loosen hard carbon deposits. Rinse parts in clean diesel fuel or calibrating oil immediately after cleaning to neutralize the solvent and prevent etching of polished surfaces. Clean nozzle spray hole orifice using cleaning wire in a pin vise. Spray hole diameter is 1 mm (0.039 inch) on Models 1120, 1220, 1320, 1520 and 1720; 0.25 mm (0.010 inch) on Model 1920; 0.24 mm (0.0095 inch) on Model 2120.

Inspect lapped surfaces of nozzle valve, spacer and nozzle holder for scratches, burrs or other defects and renew as necessary. Examine nozzle valve seat and needle for scoring, scratches or evidence of overheating. Nozzle valve body and needle are a lapped fit and may not be interchanged or renewed individually. With nozzle body held vertically and wet with diesel fuel, nozzle needle should slide freely onto its seat under its own weight. If needle sticks, reclean or renew nozzle valve assembly.

Reclean all parts by rinsing thoroughly in clean diesel fuel or calibrating oil and assemble while parts are wet with clean diesel fuel. On Models 1120, 1220, 1320, 1520 and 1720, tighten nozzle cap nut to a torque of 61-75 N·m (45-55 ft.-lbs.). On Models 1920 and 2120, tighten nozzle nut to a torque of 41-47 N·m (30-35 ft.-lbs.). Do not overtighten as distortion may cause nozzle valve to stick. Retest assembled injector and adjust opening pressure as previously outlined.

GLOW PLUGS

All Models

68. Glow plugs are parallel connected with each glow plug grounding through mounting threads in cylinder head. When key switch is turned to "HEAT" position, glow plugs should be energized and "COLD START AID" indicator light should come on. Glow plugs should be fully heated in approximately 4-5 seconds and indicator light should go out. Refer to paragraph 83 for service procedures covering glow plug electrical circuit.

COOLING SYSTEM

All models use a pressurized cooling system which raises the coolant boiling point. An impeller type centrifugal pump is used to provide forced circulation. A thermostat is used to stabilize operating temperature.

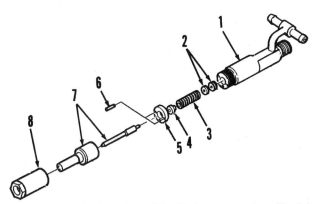

Fig. 109—Exploded view of fuel injector used on Models 1920 and 2120.

1. Nozzle holder
2. Shims
3. Spring
4. Spring seat
5. Spacer
6. Pin
7. Nozzle valve assy.
8. Nozzle nut

RADIATOR

All Models

70. Radiator cap pressure valve is set to open at 90 kPa (13 psi) on all models. It is recommended that a 50/50 mix of ethylene glycol base antifreeze and soft water be used for coolant. Refer to CONDENSED SERVICE DATA section for cooling system capacities.

To remove radiator, first remove radiator cap, open drain valve and drain the coolant. Remove engine side screens and side panels. Remove the battery. Remove air cleaner inlet tube. Disconnect radiator hoses. Remove radiator support braces. If equipped with hydrostatic transmission, disconnect hydraulic lines from oil cooler located in front of radiator. Remove nuts retaining radiator rubber mounts to frame and remove radiator from tractor.

To reinstall radiator, reverse the removal procedure.

THERMOSTAT

All Models

71. The thermostat is located between the water pump and cylinder head on Models 1120 and 1220. Water pump must be removed as outlined in paragraph 72 in order to remove the thermostat.

On all other models, thermostat is located in thermostat housing/coolant outlet elbow. To remove thermostat, drain coolant and disconnect radiator hose and bypass hose (if used) from thermostat housing. Unbolt and remove housing and thermostat.

On Models 1120, 1220 and 2120, thermostat should start to open at 71° C (160° F) and be fully open at 85° C (185° F). On Models 1320, 1520, 1720 and 1920, thermostat should start to open at 71° C (160° F) and be fully open at 82° C (180° F).

WATER PUMP

Models 1120-1220

72. R&R AND OVERHAUL. To remove water pump, first open drain valve and drain coolant from engine and radiator. Remove radiator as outlined in paragraph 70. Loosen alternator mounting bolts and remove fan belt. Unbolt and remove fan and pulley. Remove radiator hoses from water pump. Remove pump retaining cap screws and withdraw water pump assembly and thermostat.

To disassemble pump, press impeller shaft and bearing assembly (2—Fig. 110) forward out of impeller (5) and pump housing (3). Press impeller shaft out of fan hub (1). Drive seal assembly (4) out rear of pump housing.

Inspect all parts for excessive wear or damage and renew as necessary. Renew impeller shaft and bearing assembly if bearing feels rough when rotated.

When assembling pump, apply nonhardening sealer to outer surface of seal (4). Install new seal into housing by pressing on outer edge of seal. Install shaft and bearing assembly into pump housing by pressing on outer edge of bearing until front end of bearing is flush with front of housing (B—Fig. 111).

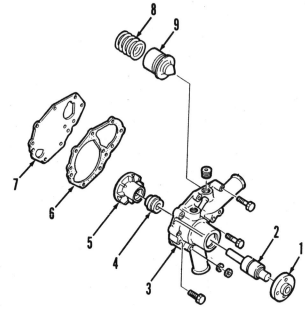

Fig. 110—Exploded view of water pump used on Models 1120 and 1220.

1. Fan hub	5. Impeller
2. Shaft & bearing assy.	6. Gasket
3. Pump housing	7. Rear plate
4. Shaft seal	8. Spring
	9. Thermostat

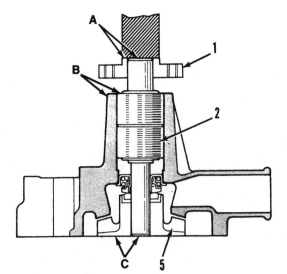

Fig. 111—When assembling water pump on Models 1120 and 1220, surfaces (A, B and C) should be flush.

Support rear end of impeller shaft, then press fan hub onto shaft until flush with front end of shaft (A). Install ceramic insert into impeller, then press impeller onto shaft until flush with rear end of shaft (C).

Reinstall water pump and thermostat using new mounting gaskets.

Models 1320-1520-1720-1920

73. REMOVE AND REINSTALL. To remove water pump, first open drain valve and drain coolant from engine and radiator. Remove radiator as outlined in paragraph 70. Loosen alternator mounting bolts and remove fan belt. Unbolt and remove fan, fan hub and pulley. Remove lower radiator hose and bypass hose from water pump. Unbolt and remove water pump and pump mounting plate from engine.

Service parts are not available for repair of water pump on these models. If coolant leaks from pump or if pump is noisy when running, renew water pump as an assembly.

To reinstall water pump, reverse the removal procedure.

Model 2120

74. R&R AND OVERHAUL. To remove water pump, first open drain valve and drain coolant from engine and radiator. Remove radiator as outlined in paragraph 70. Loosen alternator mounting bolts and remove fan belt. Unbolt and remove fan. Remove lower radiator hose and bypass hose from pump. Unbolt and remove pump and mounting plate from engine.

To disassemble pump, remove mounting plate (8—Fig. 112) from rear of pump housing. Remove bearing retaining set screw (10). Press impeller shaft and

bearing assembly (3) forward out of impeller (6) and housing. Drive seal (5) rearward from pump housing. Press impeller shaft out of fan pulley (2).

Inspect all parts for excessive wear or damage and renew as necessary. Renew impeller shaft and bearing assembly if bearing feels rough when rotated. Renew shaft seal (5).

When assembling water pump, apply nonhardening sealer to outer surface of new seal. Install seal by pressing against outer edge of seal until it bottoms against housing shoulder. Install shaft and bearing assembly into housing by pressing against outer edge of bearing until front of bearing is flush with front surface of housing. Apply several drops of silicon oil on impeller seal contact surface, then press impeller onto shaft until end of shaft is flush with rear surface of impeller. Press fan pulley onto shaft until distance (D—Fig. 113) from rear surface of pump housing to front of fan pulley is 198.3 mm (7.81 inches). Make sure pump shaft turns freely, then install bearing set screw and tighten jam nut securely.

To reinstall pump, reverse the removal procedure.

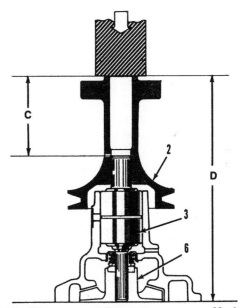

Fig. 113—When assembling water pump on Model 2120, dimension "C" should be 71.95 mm (2.830 inches) and dimension "D" should be 198.3 mm (7.810 inches).

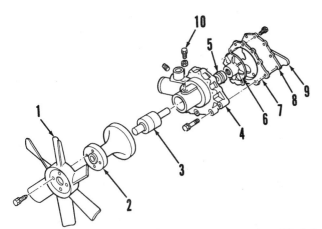

Fig. 112—Exploded view of water pump used on Model 2120.

1. Fan	
2. Fan hub	6. Impeller
3. Shaft & bearing	7. Gasket
assy.	8. Rear plate
4. Pump housing	9. Gasket
5. Shaft seal	10. Set screw

ELECTRICAL SYSTEM

ALTERNATOR AND REGULATOR

All Models

76. Models 1120 and 1220 are equipped with a Nippondenso alternator rated at 20 amperes and an external, mechanical type regulator which is mounted on the engine firewall.

Models 1320, 1520, 1720, 1920 and 2120 are equipped with a Mitsubishi alternator rated at 35 amperes and an integral, solid-state nonadjustable regulator.

When servicing the electrical system, the following precautions must be observed to avoid damage to charging circuit components.

  a. When installing battery or connecting a booster battery, the negative post must be grounded.
  b. Always disconnect battery ground cable before removing or installing any electrical components.
  c. Do not connect or disconnect any charging circuit wiring when engine is running.
  d. Never short across any terminal of alternator or regulator unless specifically recommended.

77. OUTPUT TEST. To check alternator output, disconnect output wire at alternator and connect a test ammeter in series with alternator output terminal and wire as shown in Fig. 115. Connect a load tester and voltmeter to battery terminals.

Turn on headlights for one minute, then record battery voltage. Start the engine and set engine speed at approximately 2000 rpm. Using the load tester, apply load to charging system to obtain maximum alternator output and record voltmeter and ammeter readings.

If voltmeter reading increases but remains below 15.5 volts (normal regulated voltage is 13.8-14.8 volts) and ammeter reading is minimum of 18 amperes on Models 1120 and 1220 or 25 amperes on all other models, charging system is operating properly. If voltmeter reading exceeds 15.5 volts, regulator is faulty. If voltmeter reading remains the same or decreases and ammeter reading is less than 18 or 25 amperes, alternator or regulator is defective. On Models 1120 and 1220, perform the following maximum field output test.

With engine shut off, connect a jumper wire (3—Fig. 116) between alternator output terminal (4) and field terminal (1). Start engine and perform output test using same procedure as outlined above. If voltmeter or ammeter reading remains the same or decreases, alternator is faulty. If voltmeter or ammeter reading increases, regulator or wiring is faulty. Note that there must be current at alternator "F" terminal when key switch is turned to "ON" position before alternator will begin charging.

If alternator is charging but indicator light remains "ON" when engine is running on Models 1120 and 1220, check alternator "N" circuit voltage as follows: With engine running at 1400-1800 rpm, connect voltmeter positive lead to "N" terminal (2—Fig. 116) on alternator and connect voltmeter negative terminal to ground. Voltmeter reading should be 4.0-5.8 volts. If "N" voltage is less than 4.0 volts, alternator is faulty. If "N" voltage is within specified range but light remains on, regulator is faulty.

Models 1120-1220

78. R&R AND OVERHAUL. To remove alternator, first disconnect battery ground cable. Disconnect wiring from rear of alternator. Remove mounting bolts and remove alternator.

Prior to disassembly, scribe matching marks across both end housings and stator frame for reference when assembling. Refer to Fig. 117 for an exploded view of alternator. Remove through-bolts and separate front housing (5) and rotor (11) from rear housing (17) and stator (14). Remove retaining nuts, then separate stator from rear housing. When unsoldering stator wires from diode rectifier (16), use a pre-

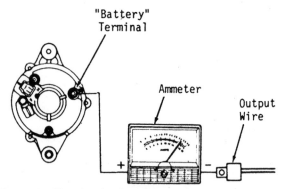

Fig. 115—To check alternator output, connect a test ammeter in series with battery terminal of alternator and the output wire.

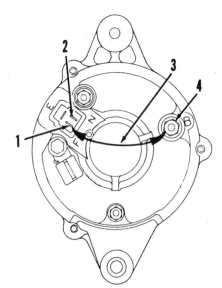

Fig. 116—To check maximum field output, connect a jumper wire (3) between "F" terminal (1) and "B" terminal (4).

heated soldering iron and separate connection as quickly as possible to avoid heat damage to diodes. Remove brush holder (13) from the rectifier. Rotor can be removed from front housing after removing retaining nut, drive pulley and fan.

Inspect all parts for wear, burned or discolored wiring or other damage and renew if necessary. Brush length when new is 12.5 mm (0.492 inch) and wear limit is 5.5 mm (0.217 inch). Using an ohmmeter, check for open, shorted or grounded circuits as follows:

Touch one tester lead to each of the rotor slip rings (3—Fig. 118). Resistance should be 4.2 ohms. Excessively low or high resistance reading indicates shorted or open circuit, and rotor should be renewed. Renew rotor if there is continuity (grounded circuit) between slip rings and rotor frame (2). Rotor slip ring diameter when new is 32 mm (1.260 inches) and wear limit is 31.7 mm (1.248 inches). Slip ring runout should not exceed 0.05 mm (0.002 inch). Slip rings may be cleaned using 400 grit silicon carbide paper and polished with crocus cloth. Do not use emery cloth.

Check stator for continuity between each of the stator wire leads (1—Fig. 119). If there is no continuity between stator leads, an open circuit is indicated. Check for continuity between each of stator leads and stator frame (2). A grounded circuit is indicated if there is continuity between any of the leads and stator frame. Renew stator if windings are discolored or burned.

Check rectifier diodes by touching one tester lead to diode terminal and touching the other lead to rectifier frame (Fig. 120), then reverse the test lead connections. The diodes should show continuity in one direction and high resistance in the other direction. If one or more diodes has high resistance in both directions or low resistance in both directions, renew rectifier assembly.

When reassembling alternator, solder stator wire connections using resin core solder. Hold rectifier terminals with needlenose pliers to absorb heat during

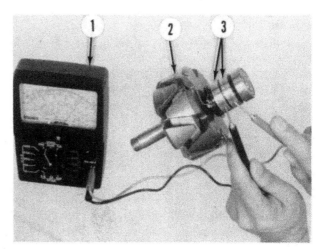

Fig. 118—Use an ohmmeter to check rotor for open or short circuits.

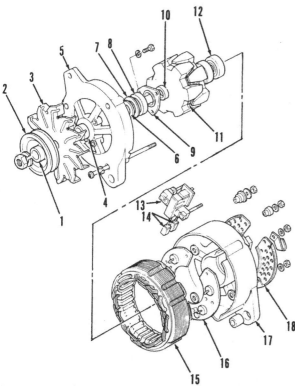

Fig. 117—Exploded view of Nippon denso alternator used on Models 1120 and 1220.

1. Spacer		11. Rotor	
2. Pulley		12. Bearing	
3. Fan		13. Regulator &	
4. Spacer		brush holder	
5. Front housing		assy.	
6. Dust seal		14. Brushes	
7. Seal cover		15. Stator	
8. Bearing		16. Rectifier assy.	
9. Retainer plate		17. Rear housing	
10. Spacer		18. Rectifier cover	

Fig. 119—Use an ohmmeter to check stator windings for open or short circuits.

soldering operation. Quickly cool soldered connection with a damp cloth to protect rectifier diodes. To hold brushes in retracted position during assembly, push brushes into holder and insert a pin or wire through holes in alternator rear cover and brush holder.

To reinstall alternator, reverse the removal procedure.

Models 1320-1520-1720-1920-2120

79. R&R AND OVERHAUL. To remove alternator, first disconnect battery ground cable. Disconnect wiring from rear of alternator. Remove mounting bolts and remove alternator.

Prior to disassembly, scribe matching marks across both end housings and stator frame for reference when assembling. Refer to Fig. 121 for an exploded view of alternator. Remove through-bolts and separate front housing (5) and rotor (9) from rear housing (16) and stator (11). Remove retaining nut, then separate stator, regulator (13) and rectifier (15) from rear housing. When unsoldering stator wires from diode rectifier (15), use a preheated soldering iron and separate connection as quickly as possible to avoid heat damage to diodes. Rotor can be removed from front housing after removing retaining nut, drive pulley and fan.

Inspect all parts for wear, burned or discolored wiring or other damage and renew if necessary. Brush length when new is 18 mm (0.71 inch) and wear limit is 8 mm (0.315 inch). Using an ohmmeter, check for open, shorted or grounded circuits as follows:

Touch one tester lead to each of the rotor slip rings (3—Fig. 118). Resistance should be 3-4 ohms. Excessively low or high resistance reading indicates shorted or open circuit, and rotor should be renewed. Renew rotor if there is continuity (grounded circuit) between slip rings and rotor frame (2). Rotor slip ring diameter when new is 33 mm (1.299 inches) and wear limit is 32.4 mm (1.276 inches). Slip rings may be cleaned using 400 grit silicon carbide paper and polished with crocus cloth. Do not use emery cloth.

Check stator for continuity between each of the stator wire leads (1—Fig. 119). If there is no continuity between stator leads, an open circuit is indicated. Check for continuity between each of stator leads and stator frame (2). A grounded circuit is indicated if there is continuity between any of the leads and stator frame. Renew stator if windings are discolored or burned.

Check rectifier diodes by touching one tester lead to diode positive heat sink (2—Fig. 122) and touching the other lead to each stator coil lead terminal (4), then reverse test lead connections. Repeat the tests at negative heat sink (3) and each stator coil lead terminal. Check the three small diodes of diode trio in the same manner. The diodes should show continuity in one direction and high resistance in the other direction. If one or more diodes has high resistance on both directions or low resistance in both directions, renew rectifier assembly.

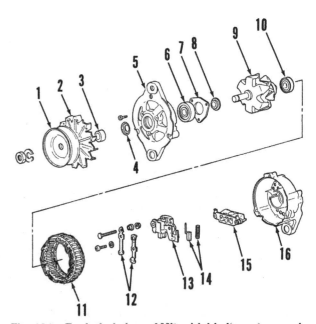

Fig. 121—Exploded view of Mitsubishi alternator used on Models 1320, 1520, 1720, 1920 and 2120.

1. Pulley	10. Bearing
2. Fan	11. Stator
3. Spacer	12. Terminals
4. Dust seal	13. Regulator &
5. Front housing	brush holder
6. Bearing	assy.
7. Retainer plate	14. Brush & spring
8. Dust seal	15. Rectifier assy.
9. Rotor	16. Rear housing

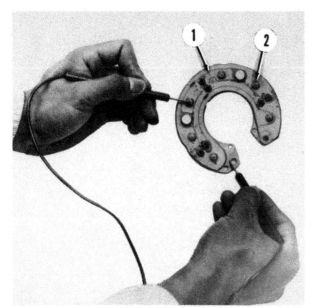

Fig. 120—Use an ohmmeter to check for faulty rectifier diodes.

When reassembling alternator, solder stator wire connections using resin core solder. Hold rectifier terminals with needlenose pliers to absorb heat during soldering operation. Quickly cool soldered connection with a damp cloth to protect rectifier diodes. To hold brushes in retracted position during assembly, push brushes into holder and insert a pin or wire through holes in alternator rear cover and brush holder.

To reinstall alternator, reverse the removal procedure.

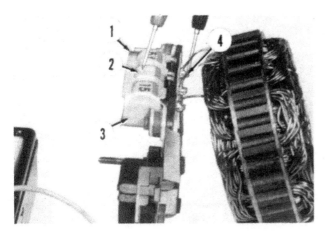

Fig. 122—Use an ohmmeter to check for faulty rectifier diodes.

1. Rectifier assy.
2. Positive heat sink
3. Negative heat sink
4. Stator coil leads

SAFETY START SWITCH

All Models

80. All tractors are equipped with two safety start switches. On Models 1120, 1220, 1320 and 1520, the starting circuit can be activated only when clutch pedal is fully depressed and pto control lever is in neutral position. The safety start switches are both located on left side of tractor (Fig. 123).

On Models 1720, 1920 and 2120, the transmission main shift lever and pto control lever must be in neutral position before starter can be activated. On these tractors, the transmission neutral start switch (1—Fig. 124) is located on the steering column, and removal of instrument panel is necessary to gain access to switch. On Models 1720 and 1920, the pto neutral start switch is located on top of rear axle center housing. On Model 2120, pto neutral start switch is located on left side of rear axle center housing.

To check and adjust safety start switches, disconnect wiring to switches and connect an ohmmeter or continuity light across switch terminals. On Models 1120, 1220, 1320 and 1520, fully depress clutch pedal and move pto control lever to neutral position. Both switches should indicate continuity across switch terminals. If not, move switch until ohmmeter or continuity light indicates switch is closed.

On Models 1720, 1920 and 2120, move transmission main shift lever and pto control lever to neutral position. Ohmmeter or continuity light should indicate continuity across terminals of both switches. The

Fig. 123—View of safety start switches on Models 1320 and 1520. Models 1120 and 1220 are similar.

1. Clutch linkage switch
2. Rear pto neutral start switch
3. Mid-pto neutral start switch

transmission neutral switch (1—Fig. 124) is a rotary type switch that has a small pin that fits into a groove in upper end of shift lever shaft (3). To adjust switch, loosen switch retaining screws (2) and move switch until continuity is indicated at switch terminals.

> **NOTE: If shift lever shaft has been removed, it may be necessary to loosen clamp bolt (5) and reposition shift lever shaft in order to correctly adjust safety switch.**

The pto neutral start switch is adjusted by means of shims located between the switch and rear axle center housing.

STARTING MOTOR AND SOLENOID

Models 1120-1220

81. OVERHAUL. The starter solenoid is enclosed in housing (9—Fig. 126). To check solenoid, use a 6 volt battery and connect wires (1, 2 and 3—Fig. 125) to locations indicated. The drive pinion should extend from the housing. Disconnect wire (3) from terminal and pinion should remain extended. If pinion does not extend with all three jumper wires connected or if pinion does not remain extended when wire (3) is disconnected, renew solenoid assembly.

To disassemble starting motor, disconnect connecting lead from solenoid. Remove through-bolts (17—Fig. 126). Unbolt and remove end cover (16). Remove brushes from brush holder (15), then remove brush holder, field coil housing (13) and armature (11) from starter housing. Remove two retaining screws and separate end frame (1) from solenoid coil housing (9). Remove starter drive/clutch assembly (2) from housing.

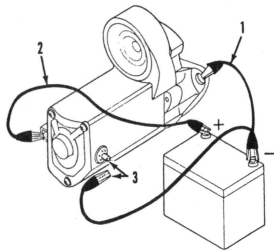

Fig. 125—Refer to text for checking starter solenoid pull-in and hold-in coils on Models 1120 and 1220.

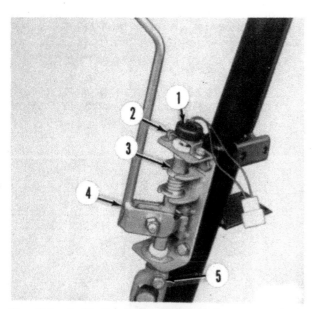

Fig. 124—On Models 1720, 1920 and 2120, transmission neutral start switch (1) is mounted on the steering column. Pto neutral start switch is mounted on top of rear axle center housing on Models 1720 and 1920 and on left side of rear axle center housing on Model 2120.

1. Transmission safety start switch
2. Retaining cap screws
3. Shift lever upper shaft
4. Shift lever
5. Clamp bolt

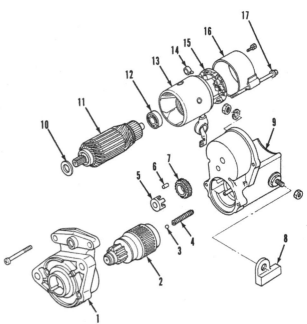

Fig. 126—Exploded view of starter motor assembly used on Models 1120 and 1220.

1. Drive end frame	9. Solenoid assy.
2. Drive clutch assy.	10. Thrust washer
3. Ball	11. Armature
4. Spring	12. Bearing
5. Retainer	13. Field coil housing
6. Roller	14. Brush spring
7. Pinion gear	15. Brush holder
8. Cover	16. End cover
	17. Through-bolt

NOTE: Do not disassemble solenoid coil and do not loosen nut on solenoid coil terminal bolt. Renew solenoid coil and housing as an assembly if faulty.

Clean all parts and check for excessive wear or other damage and renew as necessary. Armature commutator may be cleaned using sandpaper. Do not use emery cloth. Refer to the following specifications:

Armature shaft runout
  Maximum . . . . .0.05 mm
                                 (0.002 in.)
Commutator diameter
  Minimum . . . . .29 mm
                                 (1.142 in.)
Commutator runout
  Maximum . . . . .0.2 mm
                                 (0.008 in.)
Commutator insulation depth
  Minimum . . . . .0.2 mm
                                 (0.008 in.)
Brush length
  New . . . . .13.5 mm
                                 (0.531 in.)
  Wear limit . . . . .9 mm
                                 (0.354 in.)
Brush spring tension . . . . .1.2-1.7 kg
                                 (2.7-3.8 lbs.)
No-load bench test
  Current draw . . . . .90 amps
  Rpm 3000
Load test current draw . . . . .300 amps

To reassemble starter, reverse the disassembly procedure.

Models 1320-1520-1720-1920-2120

82. OVERHAUL. To disassemble starter motor, disconnect field coil strap from solenoid. Unbolt and remove solenoid (2—Fig. 127) from front housing. Remove through-bolts (24). Remove retaining screws and remove end cover (23). Remove positive brushes from brush holder (20), then remove brush holder. Withdraw armature (21) and field coil housing (19). Remove cover (18), "C" ring (17) and thrust washer (16). Unbolt and remove center bracket (15) and thrust washer (14). Push stop ring (5) inward against spring pressure to expose snap ring (4), then remove snap ring, stop ring, pinion gear and spring. Remove spring guide (11), spring (10) and lever assembly (9). Separate reduction gear (13) and pinion shaft (12) from front housing.

Clean all parts and inspect for excessive wear or other damage and renew as necessary. Armature commutator may be cleaned with fine sandpaper. Do not use emery cloth. Refer to the following specifications:

Armature shaft runout
  Maximum . . . . .0.05 mm
                                 (0.002 in.)
Commutator diameter
  Minimum . . . . .32 mm
                                   (1.260 in.)
Commutator runout
  Maximum . . . . .0.05 mm
                                   (0.002 in.)
Commutator insulation depth
  Minimum . . . . .0.2 mm
                                   (0.008 in.)
Brush length
  New . . . . .18 mm
                                 (0.710 in.)
  Wear limit . . . . .11 mm
                                 (0.433 in.)
Brush spring tension . . . . .1.3-2.2 kg
                                 (2.9-4.8 lbs.)

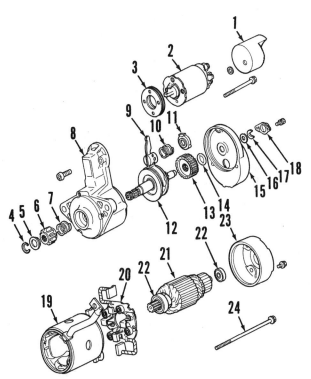

Fig. 127—Exploded view of starter motor assembly used on Models 1320, 1520, 1720, 1920 and 2120.

1. Cover
2. Solenoid
3. Shims
4. Retaining ring
5. Stop ring
6. Pinion gear
7. Spring
8. Drive end frame
9. Shift lever
10. Spring
11. Guide
12. Pinion shaft
13. Drive gear
14. Washer
15. Center housing
16. Shim
17. "C" ring
18. Cover
19. Field coil assy.
20. Brush holder assy.
21. Armature
22. Bearings
23. End cover
24. Through-bolt

No-load bench test
  Current draw . . . . .130 amps
  Rpm 4000
Load test current draw . . . . .300 amps

To reassemble starter, reverse the disassembly procedure. Check and adjust pinion shaft clearance as follows: Disconnect field coil wire from solenoid terminal. Connect a battery to solenoid and starter as shown in Fig. 128 to shift solenoid into cranking position. Push pinion gear back to remove free play and measure pinion end clearance using a feeler gage. Specified clearance is 0.5-2.0 mm (0.020-0.078 inch). Add or subtract shims (3—Fig. 127) located between solenoid and end frame as required to obtain desired clearance. Adding shims decreases clearance.

GLOW PLUGS

All Models

83. The glow plug circuit consists of fast heating type glow plugs, an indicator light and an indicator light timer unit. Glow plugs are parallel connected, with each glow plug grounding through its mounting threads. The key switch supplies electrical power to glow plugs and to indicator light and timer unit through two separate circuits (Fig. 130). When switch is turned to "Heat" position, the indicator light will come on to indicate that glow plugs are heating. In approximately 4-5 seconds, indicator light should go out, indicating that glow plugs are sufficiently heated. If key switch is inoperative in "Heat" position, glow plugs, indicator light and timer all will fail to operate. If timer unit is faulty, the indicator light will fail to illuminate, but glow plugs will operate normally.

To check key switch, remove steering column center panel. Remove retaining nut from switch and withdraw switch from steering column shroud. Connect one test lead of an ohmmeter to terminal marked 19 and connect other test lead to terminal marked 30. There should be continuity between the two terminals when key is turned to "Heat" position. If not, renew key switch.

To check glow plug indicator light and timer unit, remove instrument panel retaining screws, raise instrument panel and remove circuit board cover. Use a 12 volt test light to check for current at indicator light connector strip when key switch is turned to "Heat" position. Test light should illuminate for 4-5 seconds and then go out. If there is current at test lamp but indicator light does not come on, renew indicator lamp. If there is no current at test light, check wiring harness for broken wire or loose connection. If wiring harness is not faulty, renew timer unit. Timer unit (1—Fig. 131) is mounted on engine firewall.

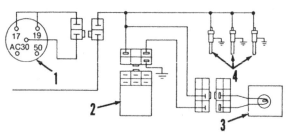

Fig. 130—Glow plug wiring circuit.
  1. Key switch
  2. Timer relay switch
  3. Indicator lamp
  4. Glow plugs

Fig. 131—Glow plug timer relay (2) is mounted on the firewall.
  1. Glow plugs
  2. Timer relay switch
  3. Fuel stop solenoid

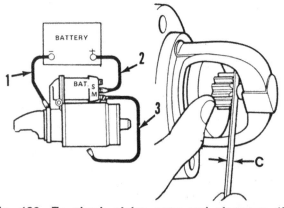

Fig. 128—To check pinion gear end clearance (C), disconnect field coil wire from solenoid terminal. Connect jumper wires from battery negative terminal to starter frame (1), from battery positive terminal to solenoid "S" terminal (2) and from solenoid "M" terminal to starter frame (3).

To check continuity-resistance of each glow plug, remove wiring connector from glow plugs and connect one test lead of an ohmmeter to glow plug terminal and connect other test lead to glow plug body. Normal resistance is 0.5 ohm for Models 1920 and 2120 and 0.8 ohm for all other models. Renew glow plug if there is high resistance (open circuit).

ENGINE FUEL STOP SOLENOID

Models 1120-1220-1320-1520-1720-1920

84. The fuel stop solenoid (3—Fig. 131) used on these tractors is spring loaded so that when solenoid is not energized (key switch "Off"), solenoid plunger pushes injection pump control rack into shut-off position. When key switch is turned to "Start" and "On" positions, solenoid plunger retracts and allows pump control rack to move to operating position.

To check solenoid operation, disconnect wiring and unscrew solenoid from engine. When solenoid is not energized, plunger (1—Fig. 132) should protrude 25.5-26.5 mm (1.004-1.043 inches). Connect a 12 volt battery to solenoid as shown in Fig. 132, then measure protrusion of plunger in retracted position, which should be 11.5-14.5 mm (0.453-0.0.571 inch). If plunger measurements are not within specified limits, renew solenoid unit.

Model 2120

85. The fuel stop solenoid (1—Fig. 133) used on Model 2120 is connected to injection pump stop lever by a spring wire. When key switch is turned to

"Off" position, solenoid pulls stop lever rearward and shuts off the injection pump. A timer inside the solenoid holds the plunger in shut-off position for about 10 seconds, then releases the plunger and allows stop lever to return to operating position so engine can be restarted.

To check solenoid operation, remove right side screen and lower panel. Disconnect wire connector and spring wire from solenoid. Unbolt and remove solenoid and mounting bracket from engine. Connect battery negative jumper wire (1—Fig. 134) to "black" connector wire terminal. Connect battery positive jumper wire (4) to "blue" connector wire terminal. Connect positive jumper wire (2) with a switch (3) to "blue/red" connector wire terminal. When switch is closed, solenoid plunger should be drawn into solenoid body and held for 8-12 seconds. If not, renew solenoid assembly.

Fig. 133—View of fuel stop solenoid (1) used on Model 2120.

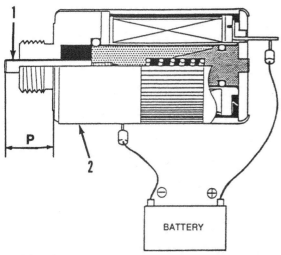

Fig. 132—To check operation of fuel stop solenoid used on all tractors except Model 2120, connect 12 volt battery to solenoid as shown and measure protrusion (P) of solenoid plunger (1). Refer to text.

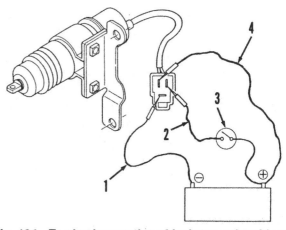

Fig. 134—To check operation of fuel stop solenoid used on Model 2120, connect 12 volt battery to solenoid wiring connector as shown. Plunger should be drawn into solenoid body for 8-12 seconds, then plunger should be released.

ENGINE CLUTCH

ADJUSTMENT

All Models

86. On all models, clutch pedal free travel should be 20-30 mm (3/4 to 1-3/16 inches) measured at pedal pad. To adjust free travel, disconnect clutch rod (2—Fig. 135) from cross shaft bellcrank (4). Loosen clevis locknut and lengthen or shorten clutch rod as required to obtain desired free travel.

On models equipped with dual plate clutch, adjustment of pto clutch release bolts (3—Fig. 136) can be checked and adjusted after removing rubber cover from side of clutch housing. Clearance between head of bolts and pto pressure plate (4) should be between 0.9-1.0 mm (0.035-0.040 inch) on Models 1320 and 1520; 1.5-1.6 mm (0.060-0.065 inch) on Model 1720; 0.95-1.0 mm (0.037-0.040 inch) on Model 1920; 1.1-1.2 mm (0.045-0.050 inch) on Model 2120.

Check adjustment of safety start switch, Models 1120, 1220, 1320 and 1520 as outlined in paragraph 80 after clutch linkage is adjusted.

TRACTOR SPLIT BETWEEN ENGINE AND CLUTCH HOUSING

All Models

87. To separate (split) tractor, drain oil from transmission housing. Disconnect ground cable from battery. Disconnect headlight wiring and remove hood, side screens and lower panels. Remove steering wheel

Fig. 135—View of clutch linkage used on Model 1920. Other models are similar. Pedal free travel should be 19 to 30 mm (3/4 to 1-3/16 inches) measured at pedal pad.

1. Clutch pedal
2. Adjustment rod
3. Clevis
4. Cross shaft
   bellcrank

using a suitable puller. Remove center panel from rear of steering column shroud, then disconnect wiring harness connectors. Disconnect tachometer cable, glow plug wire, fuel stop solenoid wire and throttle cable from injection pump. Remove retaining screws and remove instrument panel and steering column shroud. Disconnect fuel lines from fuel tank and drain the fuel. Remove fuel tank retaining strap and remove fuel tank. Disconnect wiring and remove starter motor. On Model 2120, unbolt and remove fuel stop solenoid and mounting bracket.

On models with manual steering, disconnect steering arm from steering gear shaft. On models with power steering, drain oil from power steering reservoir. Disconnect and remove power steering hydraulic tubes from steering control valve, hydraulic pump and power steering cylinder at side frame. On models equipped with front wheel drive axle, disconnect front wheel drive shaft. On models equipped with hydrostatic transmission, disconnect hydraulic lines to oil cooler located in front of radiator. On Model 2120 equipped with hydraulic shuttle shift (HSS), remove inlet tube and pressure tube from HSS hydraulic pump. On all models, disconnect and remove hydraulic system suction tube, pressure tube and oil return tube from tractor.

Place wood blocks between front axle support and engine side frame to prevent tipping. Support engine and clutch housing separately with suitable splitting stands or overhead hoist. Remove cap screws attaching engine to clutch housing, then move engine away from clutch housing.

SINGLE PLATE CLUTCH

All Models So Equipped

88. R&R AND OVERHAUL. The clutch disc (1—Fig. 137) and pressure plate assembly (2) can be removed from flywheel after separating engine from

Fig. 136—On tractors with dual plate clutch, clearance between pto clutch release bolts (3) and pressure plate (4) can be adjusted through opening in clutch housing.

clutch housing as outlined in paragraph 87. Loosen clutch mounting cap screws evenly to reduce chances of distorting pressure plate. The pressure plate is serviced as an assembly and should not be disassembled.

Check pressure plate for cracks, scoring or distortion. Inspect pressure plate release levers for wear or damage. Renew pressure plate assembly if any damage is evident. Check clutch disc for signs of overheating, distortion, loose linings, loose or broken cushioning springs or worn clutch hub splines. Disc should be renewed if depth from top of clutch lining to top of rivets is less than 0.3 mm (0.012 inch).

> **IMPORTANT: Friction surface of a new pressure plate is coated with a protective film which must be removed using a suitable solvent prior to installation.**

When installing clutch, longer side of clutch disc hub should face away from flywheel. Use a suitable pilot tool to align clutch disc while attaching clutch to flywheel. Tighten retaining cap screws evenly to torque of 23-28 N·m (17-21 ft.-lbs.). Release lever height above surface of flywheel should be 54 mm (2.13 inches) on Models 1320 and 1520 or 55 mm (2.17 inches) on Model 2120. Release lever height is not adjustable on other models.

DUAL PLATE CLUTCH

All Models So Equipped

89. R&R AND OVERHAUL. The dual plate clutch assembly can be removed from flywheel after separating engine from clutch housing as outlined in paragraph 87. Loosen clutch mounting cap screws evenly to prevent distortion of pressure plate assembly.

To disassemble clutch used on Models 1320 and 1520, use a press or other suitable means to compress clutch diaphragm springs and relieve tension on release lever links (12—Fig. 138). Then remove pins retaining release levers (11) to outer cover (8) and swing release levers and links out of the way. Release spring tension and separate clutch components.

To disassemble clutch used on Models 1720, 1920 and 2120, loosen pto clutch adjusting bolts (9—Fig. 139) evenly until bolts are free of bottom pressure plate (2). Remove pins retaining release levers (11) to outer cover (8) and separate clutch components.

Inspect pressure plates (2 and 6) for scoring, cracks or signs of overheating. Pressure plates may be resurfaced to remove minor imperfections. Check clutch discs for signs of overheating, distortion, loose linings, loose or broken cushioning springs or worn clutch hub splines. Discs should be renewed if depth

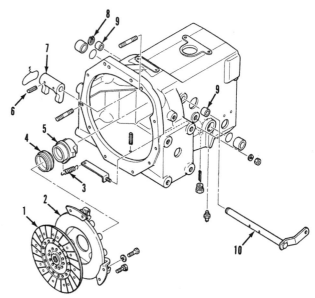

Fig. 137—Exploded view of typical single plate clutch assembly and clutch release bearing.

1. Clutch disc	6. Pin
2. Pressure plate	7. Fork
4. Release bearing	8. Snap ring
5. Release bearing	9. Bushings
hub	10. Cross shaft

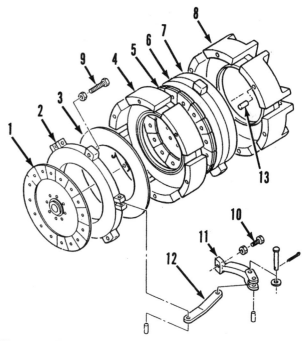

Fig. 138—Exploded view of dual plate clutch assembly available on Models 1320 and 1520.

1. Transmission	8. Clutch outer
clutch disc	cover
2. Transmission	9. Pto clutch
clutch pressure	adjusting bolt
plate	10. Release lever
3. Diaphragm spring	height adjusting
4. Clutch inner	bolt
cover	11. Clutch release
5. Pto clutch disc	lever
6. Pto pressure	12. Link
plate	13. Dowel pin
7. Diaphragm spring	

from top of clutch lining to top of rivets is less than 0.3 mm (0.012 inch).

NOTE: Friction surface of new pressure plates is coated with a protective film which must be removed using suitable solvent prior to installation.

To reassemble clutch, reverse the disassembly procedure. Be sure that dome side of diaphragm springs (3 and 7) and long side of clutch disc hubs face away from flywheel. Use alignment arbor tool number FNH 00077 on Models 1320 and 1520, tool number FNH 00078 on Models 1720 and 1920 or tool number FNH 00079 on Model 2120 to align clutch discs with flywheel. Tighten clutch mounting cap screws evenly to a torque of 34 N·m (25 ft.-lbs.).

Use a feeler gage (3—Fig. 140) to check gap between pto clutch adjusting bolt (9) and pressure plate. Adjust the bolts to obtain specified gap of 0.9-1.0 mm (0.035-0.040 inch) on Models 1320 and 1520, 1.5-1.6 mm (0.060-0.065 inch) on Model 1720, 0.95-1.0 mm (0.037-0.040 inch) on Model 1920 and 1.1-1.2 mm (0.045-0.050 inch) on Model 2120. Position release lever height setting gage (2) on the alignment arbor (1)

so that gage contacts shoulder on arbor. Adjust each release lever adjusting bolt (10) until head of each bolt just contacts bottom of adjusting gage. Height of release levers from flywheel surface should be 96.5 mm (3.80 inches) on Models 1320 and 1520, 113 mm (4.45 inches) on Model 1720, 118 mm (4.645 inches) on Model 1920 and 108 mm (4.25 inches) on Model 2120.

STANDARD 9 × 3 GEAR TRANSMISSION

(Models 1120-1220)

The standard gear transmission has three unsynchronized forward gears and one reverse gear combined with a three speed range gearbox to provide total of nine forward speeds and three reverse speeds.

LUBRICATION

Models 1120-1220

90. The rear transmission housing contains the transmission gears, differential, ring gear and pinion and serves as a common oil reservoir for lubrication of gears and bearings and oil for hydraulic system. Oil level dipstick is located in shift cover and oil filler opening is located in rear of hydraulic lift cover. It is recommended that oil be drained and refilled with Ford 134 lubricant or equivalent every 300 hours of operation.

The hydraulic system oil filter is located on right side of transmission housing. The spin-on type filter should be renewed every 300 hours of operation.

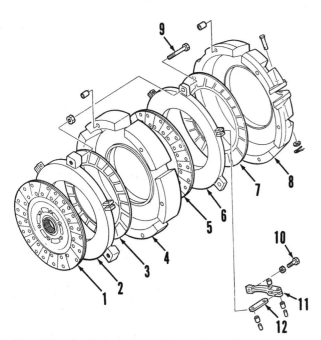

Fig. 139—Exploded view of dual plate clutch assembly used on Models 1720, 1920 and 2120.

1. Pto clutch disc
2. Pto pressure plate
3. Diaphragm spring
4. Clutch inner cover
5. Transmission clutch disc
6. Transmission clutch pressure plate
7. Diaphragm spring
8. Clutch outer cover
9. Pto clutch adjusting bolt
10. Release lever height adjusting bolt
11. Clutch release lever
12. Link

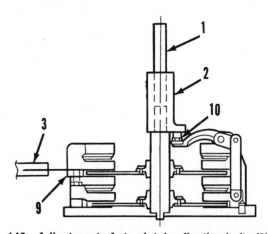

Fig. 140—Adjustment of pto clutch adjusting bolts (9) and release lever height adjusting bolts (10) are made after clutch assembly is installed on flywheel using special alignment arbor (1) and setting gage (2). Refer to text.

REMOVE AND REINSTALL
TRANSMISSION

Models 1120-1220

91. To remove transmission, first drain oil from transmission housing. Remove operator's seat and track assembly. Remove hand grips from hydraulic lift control lever, transmission range shift lever and pto levers. Unbolt and remove seat support platform, side panels, shift rod cover, ROPS frame, fenders and step plates. Unscrew shift rod turnbuckles (6—Fig. 141) to disconnect shift rods from shift cover (5). Shift rod turnbuckles have left and right hand threads. Unbolt and remove shift cover assembly. Disconnect oil pressure tube (4), oil suction tube (2) and oil return tube (3). If equipped with front wheel drive, disconnect front wheel drive shaft.

Support transmission housing and clutch housing separately with suitable support stand and hoist. Place wood blocks between front axle and side frame to prevent tipping. Remove cap screws retaining transmission housing to clutch housing. Note that two cap screws are located inside transmission housing

and are accessible through shift cover opening. Roll transmission away from clutch housing.

To reinstall transmission, reverse the removal procedure.

OVERHAUL

Models 1120-1220

92. To disassemble transmission, disconnect pto control linkage and range gear shift lever. Remove hydraulic lift control lever. Remove hydraulic valve from side of lift housing. Disconnect lift links from lift arms. Remove top link and top link bracket from rear of lift housing. Unbolt and remove hydraulic lift cover from transmission housing.

Remove retaining screw (54—Fig. 143) and withdraw transmission input shaft (55) from rear of clutch housing. Remove idler gear (61) and shaft (56) from front of transmission housing. Remove snap ring (24) and withdraw countershaft drive gear (25). Unbolt and remove bearing retainer plate (26).

Remove snap rings from front and rear of mainshaft (16). Disengage snap rings (13) from grooves on mainshaft, then drive mainshaft forward until rear bearing (22) can be removed. Slide gears (19 and 21) and spacer (20) off rear of shaft. Drive the shaft rearward and remove front bearing (12), sliding gears (14 and 15) and snap rings (13). Remove mainshaft from front of housing while removing center bearing (17) from rear of shaft.

If equipped with front wheel drive, remove hydraulic filter base (6—Fig. 144) and inlet tube (4) from side of transmission housing. Remove front wheel drive shift linkage (1 and 2) and cover (3). Remove snap ring

Fig. 141—View of disconnect points when splitting transmission housing from clutch housing on Models 1120 and 1220. Shift rod turnbuckles (6) have left and right hand threads.

1. Hydraulic filter
2. Oil suction tube
3. Oil return tube
4. Oil pressure tube
5. Transmission shift cover
6. Shift rods

Fig. 142—View of Models 1120 and 1220 standard 9 × 3 transmission with shift cover and hydraulic lift cover removed.

3. Front wheel drive shaft	21. High range gear
14. First & third sliding gear	23. Snap ring
15. Second & reverse sliding gear	30. Front countershaft
16. Mainshaft	41. Rear countershaft & bevel drive gear
19. Low range gear	

(10—Fig. 143) and sliding gear (9) from rear of front wheel drive shaft (3). Disengage snap rings (4) from grooves of shaft, then pull shaft forward from transmission housing while removing thrust washers (5), drive gear (6) and bearing (7). Remove shaft oil seal (1) from rear of clutch housing.

Disengage rear snap ring (28—Fig. 143) from groove in countershaft (30). Slide countershaft forward and remove front bearing (27). Lift countershaft with gear (29) out top of transmission housing. Remove snap ring (32) and needle bearing (31) from rear of shaft.

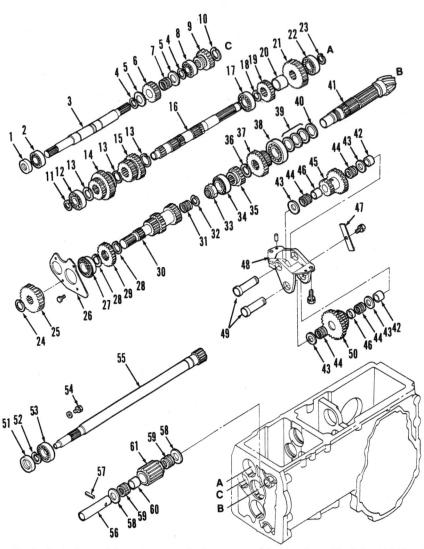

Fig. 143—Exploded view of 9 × 3 transmission assembly used on Models 1120 and 1220.

1. Seal	16. Mainshaft	32. Snap ring
2. Bearing	17. Bearing	33. Locknut
3. Front wheel	18. Snap ring	34. Bearing
drive shaft	19. Low range gear	35. Front wheel drive
4. Snap rings	20. Spacer	gear
5. Thrust washers	21. High range gear	36. Snap ring
6. Drive gear	22. Bearing	37. Range sliding gear
7. Needle bearing	23. Snap ring	38. Bearing
8. Bearing	24. Snap ring	39. Shims
9. Sliding gear	25. Drive gear	40. Thrust washer
10. Snap ring	26. Retainer plate	41. Rear countershaft
11. Snap ring	27. Bearing	& bevel drive gear
12. Bearing	28. Snap ring	42. Spacer
13. Snap rings	29. Gear	43. Thrust washer
14. Sliding gear, 1st	30. Front	44. Needle bearing
& 3rd	countershaft	45. Mid-range gear
15. Sliding gear, 2nd	31. Needle bearing	cluster
& rev.		

46. Spacer
47. Retainer plate
48. Mid-range housing
49. Shafts
50. Mid-range gear
cluster
51. Seal
52. Snap ring
53. Bearing
54. Cap screw
55. Transmission input
shaft
56. Idler shaft
57. Pin
58. Thrust washer
59. Needle bearing
60. Spacer
61. Idler gear

Refer to paragraphs 130 through 133 for service procedures outlining removal, installation and adjustment of bevel pinion shaft (41).

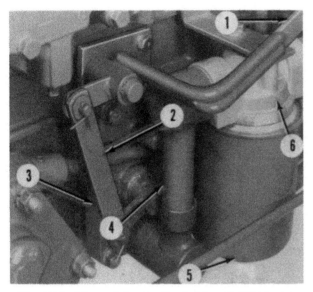

Fig. 144—View of front wheel drive shift linkage.

1. Shift lever	4. Oil suction tube
2. Shift link	5. Hydraulic oil filter
3. Shifter cover	6. Oil filter base

To remove pto front drive shaft (3—Fig. 145), first unbolt and remove pto output shaft bearing retainer (17), pto output shaft (19) and rear drive shaft (16) from transmission housing. Disengage front snap ring (4) from groove in pto front drive shaft, then drive the shaft rearward and remove front bearing (1), sliding gear (2), snap ring (4), thrust washers (5), drive gear (7) and needle bearings (6). Tap shaft with rear bearing forward out of housing.

If equipped with mid-mount pto, disengage snap rings (12—Fig. 145) from grooves of idler shaft (11). Slide shaft forward from transmission housing while removing thrust washers (13) and idler gear (15).

The mid-range gear clusters (45 and 50—Fig. 143) are located in a housing which is mounted on underside of hydraulic lift cover. To disassemble, unbolt and remove retaining plate (47). Slide shafts (49) out of housing while removing thrust washers (43), spacers (42 and 46), gears (45 and 50) and bearings (44). Unbolt and remove housing (48) from lift cover if necessary.

To reassemble transmission, reverse the disassembly procedure while observing the following points:

When installing pto front drive shaft (3—Fig. 145), install rear bearing and snap ring on the shaft first, then install shaft and bearing into the housing from

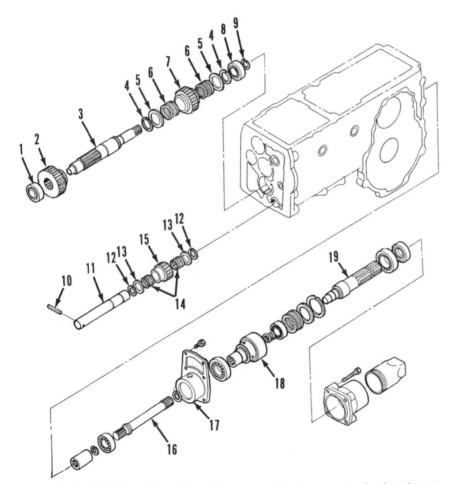

1. Bearing
2. Sliding gear
3. Pto front drive shaft
4. Snap ring
5. Thrust washer
6. Needle bearing
7. Idler gear
8. Bearing
9. Snap ring
10. Pin
11. Idler shaft
12. Snap ring
13. Thrust washer
14. Needle bearing
15. Mid-pto gear
16. Pto intermediate drive shaft
17. Bearing retainer
18. One-way clutch
19. Pto output shaft

Fig. 145—Exploded view of pto drive shaft, gears and related parts used on Models 1120 and 1220.

the front. Drive the shaft rearward far enough to permit installation of snap ring, thrust washers, bearings and gear. Install front bearing (1) flush with front of transmission housing and make sure that shoulder on front of shaft is against inner race of the bearing. Reinstall intermediate pto drive shaft and output shaft assembly, making sure all shafts turn freely.

When installing countershaft (30—Fig. 143), position snap rings (28) and drive gear (29) on shaft before installing the shaft assembly into housing from the top.

Insert front wheel drive shaft (3) into housing from the front while assembling thrust washers, gears and bearings onto shaft. Install front bearing (2) flush with front face of housing.

Position mainshaft (16) in housing as far to the rear as possible. Assemble center bearing (17), snap rings (13) and sliding gears (14 and 15) onto front of shaft, then slide shaft forward in housing. Install snap ring (18), small gear (17), spacer (20) and large gear (21) on rear of shaft. Install rear bearing (22) into housing and onto shaft from the rear and secure with snap ring (23). Drive center bearing (17) into bore in housing, then slide snap ring (18) into groove in shaft. Install bearing (12) and snap rings (11 and 13) onto front of shaft.

Install bearing retainer plate (26), countershaft drive gear (25) and idler gear (61) and shaft (56).

STANDARD 9 × 3 GEAR TRANSMISSION

(Models 1320-1520)

The standard gear transmission has three unsynchronized forward gears and one reverse gear combined with a three speed range gearbox to provide total of nine forward speeds and three reverse speeds. The main transmission gears are located in front transmission housing and range gears are located in rear axle center housing.

LUBRICATION

Models 1320-1520

93. The transmission housing and rear axle center housing serve as a common oil reservoir for lubrication of gears and bearings and oil for hydraulic system. Oil level dipstick is located on transmission top cover and oil filler opening is located in rear of hydraulic lift cover. It is recommended that oil be drained and refilled with Ford 134 lubricant or equivalent every 300 hours of operation.

The hydraulic system oil filter is located on right side of transmission housing. The spin-on type filter should be renewed every 300 hours of operation.

REMOVE AND REINSTALL TRANSMISSION

94. To remove transmission, first separate engine from clutch housing as follows: Disconnect battery ground cable. Remove hood, side screens and lower panels. Remove steering wheel. Remove center panel from steering column shroud, then disconnect wiring harness connectors. Disconnect tachometer cable and throttle control cables. Remove retaining screws and remove instrument panel and steering column shroud assembly. Disconnect fuel supply line and fuel return lines. Drain the fuel tank and remove tank from tractor. Remove starter motor. Remove hydraulic suction tube, pressure tube and oil return tube. Drain oil from power steering reservoir, then remove hydraulic lines from power steering control valve. Place wooden wedges between front axle and engine side rails to prevent tipping. Support transmission with suitable stands. Support engine with rolling jack or hoist. Remove cap screws retaining engine to clutch housing, then roll engine away from clutch housing.

Drain oil from transmission and rear axle center housing. Remove seat and seat platform. Remove ROPS bar, fenders and step plates. Disconnect rear wiring harness. Remove brake pedal control rods from both sides. Remove cap screws retaining clutch housing to transmission housing. Slide clutch housing forward while rotating shift lever to separate shift lever ball end (1—Fig. 147) from shift rail notches (2).

Fig. 147—When removing clutch housing from transmission housing, rotate shift lever to separate shift arm ball end (1) from notches in shift rails (2).

Remove top cover from transmission housing and rear axle center housing. Support transmission housing and rear axle center housing, then remove retaining cap screws and slide transmission forward from rear housing.

To reinstall transmission, reverse the removal procedure.

OVERHAUL

Models 1320-1520

95. To disassemble, proceed as follows: To remove transmission input shaft (16—Fig. 149) and related

components on models equipped with dual plate clutch, unbolt and remove bearing retainer housing (9) with input shaft and gears as an assembly from rear of clutch housing. Remove snap ring (15) and withdraw input shaft (16) from retainer housing. Remove gear (13) and bearing (14). Remove snap ring (12), then tap shaft and gear (10) out of bearing (11). Withdraw pto input shaft (3) from clutch housing.

To remove transmission input shaft (16—Fig. 150) on models equipped with single disc clutch, remove snap ring (20) and gear (19) from rear of input shaft. Disconnect clutch release bearing return spring from input shaft front bearing retainer. Remove retainer cap screws and withdraw bearing retainer and input shaft rearward from clutch housing. Remove snap ring (18) and tap input shaft and bearing out of retainer. Drive out oil seal (15).

To remove main transmission gear shift rods and forks, first remove upper detent spring and ball (4—Fig. 148) from top of transmission housing. Drive roll pin out of second-reverse shift fork (16), then pull top shift rod (28) from housing. Remove shift fork and

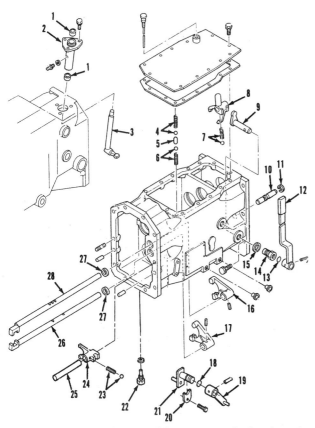

Fig. 148—Exploded view of front transmission housing and shift components.

1. Bushings	15. Seal
2. Shift lever guide	16. Shift fork, 2nd-rev.
3. Shift arm	17. Shift fork, 1st-3rd
4. Detent ball & spring	18. "O" ring
5. Interlock pin	19. Pto shift lever
6. Detent ball & spring	20. Retainer plate
7. Detent ball & spring	21. Shift arm
8. Front wheel drive shift fork	22. Bolt
9. Shift arm	23. Detent ball & spring
10. Shift shaft	24. Pto shift fork
11. Snap ring	25. Shift rod
12. Shift lever	26. Shift rod, 1st-3rd
13. "O" ring	27. Seals
14. Guide	28. Shift rod, 2nd-rev.

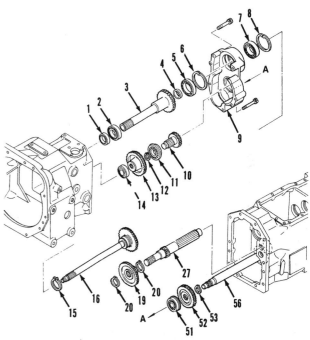

Fig. 149—Exploded view of transmission and pto input shafts and related components used on Models 1320 and 1520 equipped with 9 × 3 transmission and dual disc engine clutch.

1. Oil seal	13. Gear
2. Bearing	14. Bearing
3. Pto input shaft	15. Snap ring
4. Seal	16. Transmission input shaft
5. Bearing	19. Gear
6. Snap ring	20. Snap rings
7. Bearing	27. Transmission countershaft
8. Snap ring	51. Bearing
9. Housing	52. Gear
10. Gear & shaft assy.	53. Snap ring
11. Bearing	56. Pto drive shaft
12. Snap ring	

interlock pin (5). Drive roll pin out of first-third shift fork (17), then pull lower shift rod (26) from housing being careful not to lose lower detent ball and spring (6).

To disassemble main transmission, remove snap ring (1—Fig. 150) retaining mainshaft front bearing (2). Disengage snap rings (7) from grooves in mainshaft (4) and slide them rearward on the shaft. Drive mainshaft forward from housing while removing gears and snap rings from rear of shaft.

Remove snap ring (35—Fig. 150) from front of housing. Drive reverse idler shaft (38) forward from housing and withdraw reverse idler gear (39) out top of housing.

Pull bearing (22—Fig. 150) and gear (23) off front of countershaft (27). Remove snap ring (25) from front of transmission housing, then drive countershaft and front bearing (26) forward and remove rear bearing (33), gears (28, 29 and 31) and spacers (30 and 32) from rear of shaft.

To remove pto countershaft (56—Fig. 151), first remove snap ring (61), thrust washers (57), gears (60),

bearings (58) and spacer (59) if equipped with front wheel drive. On all models, remove snap ring (55) retaining center bearing (54). Drive the shaft rearward and remove bearing (51) and gear (52) from front of shaft.

If equipped with front wheel drive axle, drive roll pin out of shift lever (12—Fig. 148) and remove the lever. Pull shift rod (10) from rear of housing and remove shift fork (8) and shifter arm (9). Drive front wheel drive shaft (47—Fig. 151) out front of housing while removing sliding gear (48) and rear bearing (49).

To disassemble range transmission, disengage rear snap ring (3—Fig. 153) from groove of range mainshaft (2) and slide it rearward on the shaft. Drive mainshaft forward from rear housing while sliding gears (4 and 6) and spacer (5) off rear of shaft.

Drive roll pin out of range shift lever (8—Fig. 152) and remove the lever. Pull range shift rod (6) out front of housing, being careful not to lose detent ball and spring (4) as rod is removed from shift fork (3). Remove retaining plate (7) and withdraw shift arm (1).

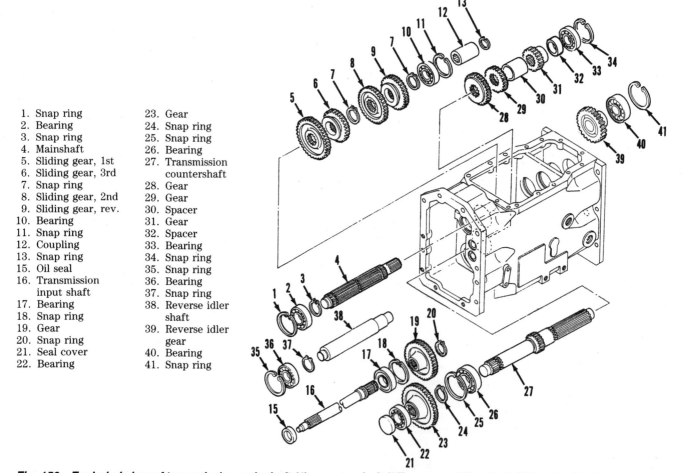

1. Snap ring	23. Gear
2. Bearing	24. Snap ring
3. Snap ring	25. Snap ring
4. Mainshaft	26. Bearing
5. Sliding gear, 1st	27. Transmission
6. Sliding gear, 3rd	countershaft
7. Snap ring	28. Gear
8. Sliding gear, 2nd	29. Gear
9. Sliding gear, rev.	30. Spacer
10. Bearing	31. Gear
11. Snap ring	32. Spacer
12. Coupling	33. Bearing
13. Snap ring	34. Snap ring
15. Oil seal	35. Snap ring
16. Transmission	36. Bearing
input shaft	37. Snap ring
17. Bearing	38. Reverse idler
18. Snap ring	shaft
19. Gear	39. Reverse idler
20. Snap ring	gear
21. Seal cover	40. Bearing
22. Bearing	41. Snap ring

Fig. 150—Exploded view of transmission mainshaft (4), countershaft (27), reverse idler shaft (38) and related components used on Models 1320 and 1520 with 9 × 3 transmission. Transmission input shaft (16) is used on tractors with single disc engine clutch.

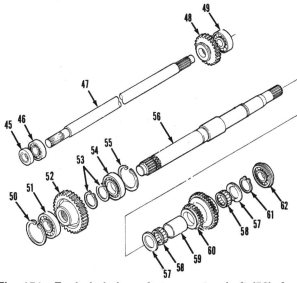

Fig. 151—Exploded view of pto countershaft (56), front wheel drive shaft (47) and related components used on Models 1320 and 1520 equipped with front wheel drive axle.

45. Oil seal	54. Bearing
46. Bearing	55. Snap ring
47. Front wheel drive shaft	56. Pto countershaft
48. Sliding gear	57. Thrust washers
49. Bearing	58. Needle bearings
50. Snap ring	59. Spacer
51. Bearing	60. Gear cluster
52. Drive gear	61. Snap ring
53. Snap ring	62. Bearing

Removal, installation and adjustment of bevel pinion shaft (8—Fig. 153) and related components is covered in paragraphs 135, 137, 138 and 139.

Reassemble remainder of shafts, gears, bearings and associated components in reverse order of disassembly procedure.

HYDROSTATIC TRANSMISSION

(Models 1120-1220-1320-1520)

The hydrostatic transmission consists of a variable displacement piston type hydraulic pump and a fixed displacement piston type motor located in the clutch housing. The hydrostatic transmission input shaft is

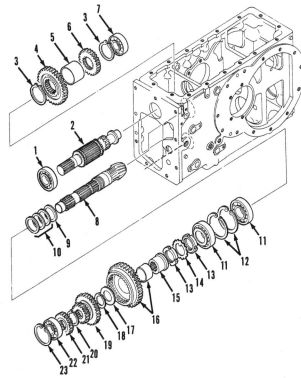

Fig. 153—Exploded view of rear mainshaft (2), bevel pinion shaft (8) and related components used on Models 1320 and 1520 with 9 × 3 transmission.

1. Bearing	
2. Rear mainshaft	14. Lockwasher
3. Snap ring	15. Coupling
4. Gear, high range	16. Gear & bushing assy.
5. Spacer	
6. Gear, mid-range	17. Thrust washer
7. Bearing	18. Snap ring
8. Bevel pinion shaft	19. Range sliding gear
9. Thrust washer	20. Snap ring
10. Shims	21. Front wheel drive gear
11. Bearing	
12. Snap ring	22. Bearing
13. Nuts	23. Snap ring

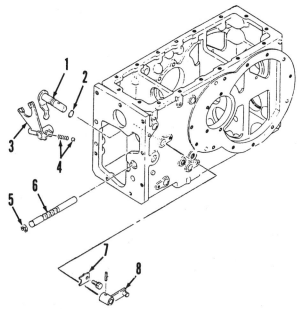

Fig. 152—Exploded view of range transmission shift fork and rod.

1. Shift arm	
2. "O" ring	5. Snap ring
3. Shift fork	6. Shift rod
4. Detent ball & spring	7. Retainer plate
	8. Shift lever

connected directly to engine clutch disc. The input shaft is splined to the hydrostatic pump, and extends out the rear of hydrostatic unit to drive the pto shaft. The output shaft is splined to hydrostatic motor and drives the two speed (Models 1120 and 1220) or three speed (Models 1320 and 1520) range transmission.

Refer to Fig. 155 for schematic drawing of hydrostatic transmission. When hydrostatic pump swash plate is in neutral (vertical) position, there is no pumping action of pump pistons and no flow of oil from pump (9) to motor (14). Oil flows from charge pump (3) through oil cooler (5) and oil filter (6) to feed valves (7 and 11) and neutral valves (8 and 10). The charge pump relief valve (12) maintains 420-560 kPa (61-81 psi) pressure in the circuit to provide lubrication for pump and motor cylinder blocks.

When pump swash plate is tilted to forward drive position, oil is discharged from piston pump (9) through "A" circuit. Pump pressure closes feed valve (7) and neutral valve (8) return to sump passages. All oil flow from piston pump then flows to the motor causing it to rotate. Oil discharged from the motor in passage "B" returns to the pump to complete the cycle. Any oil lost during the cycle due to leakage causes a drop in pressure in suction passage "B." Feed valve (11) then opens and make-up oil is provided by the charge pump through the feed valve. A high pressure relief valve (13) limits the discharge pressure of piston pump to protect the system from excessive pressure.

When pump swash plate is tilted in the opposite direction for reverse, oil flow from piston pump to

motor is reversed and passage "B" becomes the pressure passage and passage "A" is the suction passage.

LUBRICATION

All Models So Equipped

97. The hydrostatic unit utilizes fluid contained in a common reservoir consisting of transmission and differential housing. The fluid should be drained and refilled with new Ford 134 lubricant or equivalent fluid every 300 hours of operation. Capacity is approximately 16 liters (16.9 U.S. quarts) on Models 1120 and 1220 or 22 liters (23.25 U.S. quarts) on Models 1320 and 1520.

All models are equipped with a hydraulic system oil filter and a hydrostatic system oil filter. On Models 1120 and 1220, both filters are located on right side of transmission housing. On Models 1320 and 1520, hydraulic system filter is located on right side of transmission housing and hydrostatic system filter is mounted on left side of engine side rail. Both filters should be renewed every 300 hours of operation.

TROUBLE-SHOOTING

All Models So Equipped

98. The following are some symptoms which may occur during operation of the hydrostatic transmission and their possible causes.

1. Reservoir
2. Hydraulic filter
3. Charge pump
4. Input shaft
5. Oil cooler
6. Hydrostatic oil filter
7. Feed valve "A"
8. Forward neutral valve
9. Hydrostatic pump
10. Reverse neutral valve
11. Feed valve "B"
12. Charge pressure relief valve
13. High pressure relief valve
14. Hydrostatic motor
15. Output shaft

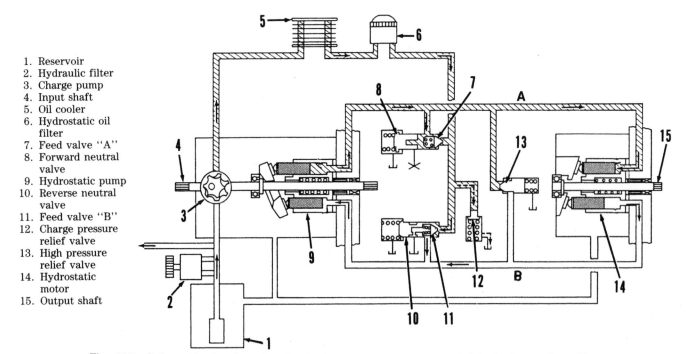

Fig. 155—Schematic drawing showing hydrostatic transmission oil flow in forward position.

1. Tractor will not move or operates erratically.
   a. Engine clutch or transmission input shaft failure.
   b. Transmission oil level low.
   c. Charge pump or charge pressure relief valve defective.
   d. Pump or motor cylinder block or valve plates defective.
   e. Feed valve or neutral valve sticking open.
2. Abnormal noise when operating.
   a. Hydrostatic oil filter plugged.
   b. Air leak on suction side of charge pump.
   c. Oil viscosity too high due to incorrect oil being used.
3. Tractor will move but lacks pulling power.
   a. High pressure relief valve opening pressure set too low.
   b. Internal leakage of high pressure oil.
4. Tractor fails to stop at neutral.
   a. Control linkage out of adjustment.
   b. Defective neutral valve.
5. Oil overheating.
   a. Oil cooler plugged.

TESTS AND ADJUSTMENTS

All Models So Equipped

99. HIGH PRESSURE RELIEF VALVE. To check high pressure relief valve pressure setting, first operate tractor until oil temperature is 40°-50° C (100°-120° F). Remove center panel from rear of steering column shroud. Remove plug from pressure test port (1—Fig. 156) and connect a 0-35000 kPa (0-5000 psi) pressure gage to test port fitting. Lock the parking brake and place range shift lever in high range. Start engine and operate at 2500 rpm. Slowly move transmission foot control pedal to obtain maximum pres-

Fig. 156—High pressure test port adapters (1) are located on top of clutch housing.

sure. Pressure reading should be 23445-25510 kPa (3400-3700 psi) on Models 1120 and 1220 or 26480-28440 kPa (3840-4125 psi) on Models 1320 and 1520.

NOTE: Do not maintain maximum pressure any longer than necessary to prevent overheating and possible damage to transmission.

If necessary to adjust the relief valve, the tractor must be separated between engine and clutch housing as outlined in paragraph 102 for access to relief valve (51—Fig. 164 or 165) adjusting screw.

100. CHARGE PUMP RELIEF VALVE. To check charge pressure relief valve setting, first operate tractor until oil temperature is 40°-50° C (100°-120° F). Remove banjo bolt (1—Fig. 157) from charge pressure supply line and install a special banjo bolt (2) that has been drilled and tapped to 1/8 inch NPT for installing test gage fitting. Connect a 0-2000 kPa (0-300 psi) test gage to banjo bolt. Place range transmission in neutral. Start engine and run at 2500 rpm. Pressure gage reading should be 420-560 kPa (61-81 psi).

Charge relief valve pressure setting is not adjustable. If pressure is low, check for plugged hydraulic filter, faulty relief valve (48—Fig. 164 or 165), faulty charge pump or faulty hydrostatic pump and/or motor cylinder block assemblies.

101. CONTROL LINKAGE ADJUSTMENT. On Models 1120 and 1220, adjust length of hydrostatic control rod (6—Fig. 158) to 145 mm (5.7 inches). Remove center panel from rear of steering column shroud and loosen cap screws retaining neutral return spring bracket (4) to steering column. Place speed control lever (1) in release position and shift range transmission to low range. Start engine and run at idle speed. Adjust neutral control rod (5) to provide sufficient tension to return foot pedal to neutral position when released. Stop the engine and tighten mounting bracket bolts. Move foot pedal to

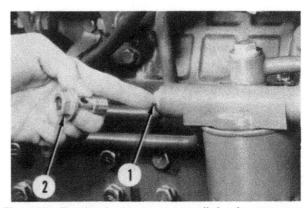

Fig. 157—To check charge pump relief valve pressure, remove banjo bolt (1) from hydrostatic oil filter outlet line and install a special banjo bolt (2) that has been drilled and tapped to allow installation of a pressure gage.

fully depressed position. Adjust pedal stop bolt to just contact the pedal, then lengthen it one more turn.

On Models 1320 and 1520, remove center panel from rear of steering column shroud. Adjust length of rod (4—Fig. 159) between foot pedal (2) and pivot linkage (1) to 129.5 mm (5.1 inches). Adjust length of rod (3) between pivot linkage and hydrostatic unit to 137 mm (5.40 inches). Start engine and shift range transmission to low range. Adjust neutral rod spring tension nut (4—Fig. 160) to provide sufficient tension to return foot pedal to neutral position when released. Stop the engine and move foot pedal to fully depressed position. Adjust length of pedal stop bolt (5—Fig. 159) to just contact step plate, then lengthen it two more turns.

REMOVE AND REINSTALL

Models 1120-1220

102. To remove hydrostatic unit, drain the transmission oil. Split the tractor between clutch housing

Fig. 159—View of hydrostatic transmission control linkage used on Models 1320 and 1520. Refer to text for adjustment procedure.

1. Pivot linkage
2. Foot control pedal
3. Hydrostatic control rod
4. Pedal control rod
5. Pedal stop bolt

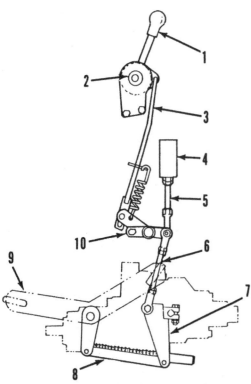

Fig. 158—Drawing of hydrostatic transmission control linkage used on Models 1120 and 1220. Refer to text for adjustment procedure.

1. Hand control lever
2. Lever friction plate assy.
3. Control rod
4. Neutral return spring assy.
5. Neutral control rod
6. Hydrostatic control rod
7. Control lever arm
8. Damper assy.
9. Foot control pedal
10. Pivot linkage

Fig. 160—View of hydrostatic hand control linkage and neutral return spring assembly.

1. Lever friction plate assy.
2. Control rod
3. Neutral return spring assy.
4. Neutral control rod
5. Pivot linkage
6. Spring

and transmission extension housing as follows: Disconnect battery ground cable. Remove steering wheel. Remove center panel from rear of steering column shroud and disconnect wiring harness connectors. Disconnect tachometer cable and throttle control cable. Unbolt and remove instrument panel and steering column shroud. Remove step plates from both sides and remove transmission cover plate. Disconnect clutch pedal rod. Remove hydrostatic pressure test port tubes from top of clutch housing. Disconnect hydrostatic control rod (15—Fig. 161). Disconnect and remove hydraulic system suction tube (5), pressure tube (3) and oil return tube (10). Remove hydrostatic filter inlet tube (7) and pressure tube (1). Remove hydrostatic pressure tube and oil return tube from left side of clutch housing. Remove brake pedals (11). If equipped with front wheel drive, disconnect drive shaft. Place wood wedges between engine side rails and front axle to prevent tipping. Support transmission and engine with suitable splitting stands and hoist or floor jack. Remove cap screws retaining clutch housing to extension housing, then roll rear axle assembly away from clutch housing.

Disconnect hydrostatic control linkage and damper assembly from hydrostatic unit. Remove hydrostatic unit retaining bolts and remove hydrostatic unit from extension housing.

To reinstall hydrostatic unit, reverse the removal procedure.

Models 1320-1520

103. To remove hydrostatic unit, first drain transmission oil. Separate clutch housing from engine as follows: Remove hood, side screens and lower panels. Disconnect battery ground cable. Drain oil from power steering reservoir and transmission housing. Remove steering wheel. Remove center panel from rear of steering column shroud. Disconnect wiring harness connectors from instrument panel, key switch and rear wiring harness. Disconnect tachometer cable and throttle control cable. Unbolt and remove instrument panel and steering column shroud. Disconnect fuel lines and drain fuel from tank. Disconnect fuel gage sender wire, remove fuel tank retaining band and remove fuel tank. Remove hydraulic system oil suction tube (2—Fig. 162), pressure tube (1), oil return tube (3) and hydrostatic transmission oil inlet tube (4). Disconnect power steering oil tubes (6) from steering control valve and hydraulic pump. Disconnect oil cooler lines (9 and 10—Fig. 163) and hydrostatic return oil line (12). Remove starter motor. Place wood wedges between engine side frame and front axle to prevent tipping. Support transmission and engine with suitable stands and overhead hoist. Remove cap screws retaining engine to clutch housing, then carefully roll engine and front end assembly away from tractor.

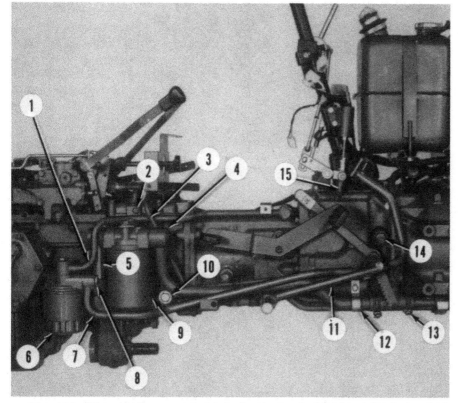

1. Hydrostatic charge pressure tube
2. Filter mounting bracket
3. Hydraulic system pressure tube
4. Oil suction tube
5. Hydraulic filter inlet tube
6. Hydrostatic oil filter
7. Filter oil inlet tube
8. Banjo bolt
9. Hydraulic system filter
10. Hydraulic oil return line
11. Brake pedal
12. Clamp
13. Hose clamp
14. Banjo bolt
15. Hydrostatic control rod

Fig. 161—View of disconnect points for removal of hydrostatic transmission unit on Models 1120 and 1220.

Remove step plates from both sides. Disconnect brake rods (13—Fig. 163). Disconnect hydrostatic control rods from control lever (11). Remove hydrostatic pressure test port tubes from top of clutch housing. Support clutch housing, remove retaining cap screws and separate clutch housing from transmission housing. Remove four retaining cap screws and stud nuts, then withdraw hydrostatic unit from front of transmission housing.

To reinstall hydrostatic unit, reverse the removal procedure.

Fig. 162—View of disconnect points for removal of hydrostatic transmission unit on Models 1320 and 1520.

1. Hydraulic system
   pressure tube
2. Oil suction tube
3. Hydraulic oil
   return tube
4. Hydrostatic inlet
   tube
5. Hose clamp
6. Power steering
   tubes
7. Power steering
   control valve

Fig. 163—View of disconnect points on left side of tractor for Models 1320 and 1520.

  8. Hose connector
  9. Oil cooler return
     tube
10. Tube to oil cooler
11. Hydrostatic
     control lever
12. Hydrostatic oil
     return tube
13. Brake rod

OVERHAUL

All Models

104. Prior to disassembling hydrostatic unit, thoroughly clean unit and plug all openings to prevent entry of dirt. Maintain a clean work area and be careful when handling components to prevent damage to precision surfaces.

Remove four socket head bolts retaining pump housing (29—Fig. 164 or 165) to port block (56). Separate pump housing from port block, being careful not to drop valve port plate (16) which will adhere to either the pump cylinder block or port block. Withdraw pump cylinder block assembly (17 through 21) from housing. Remove thrust plate (22) from swash plate (23).

Remove four socket head bolts retaining motor housing (7) to port block. Separate motor housing from port block, being careful not to drop valve port plate (15). Withdraw motor cylinder block and piston assembly (10 through 14) from motor housing. Remove thrust plate (9) from motor housing.

To remove charge pump, remove clutch release bearing hub (45) from charge pump housing (43). Unbolt and remove charge pump assembly from housing (29).

NOTE: Identify the position of inner and outer rotors (41 and 42) so that they can be reassembled in original positions.

Remove snap ring (37) from housing, then drive input shaft (24) and bearing (35) from housing. Scribe alignment marks on swash plate covers (26 and 33). Remove retaining cap screws from covers, then tap on one end of swash plate trunnion and then the other end with a soft hammer to force covers from pump housing.

Unbolt and remove end cover (1) from motor housing (7). Use a soft hammer to tap output shaft (6) and bearing (5) from housing. Drive the seal (2) out of end cover.

Remove charge pump relief valve poppet and spring (48) from port block. Remove neutral valves (53) and feed valves (55) from port block. Remove high pressure relief valve assembly (51). When removing high pressure relief valve, count the number of turns required to remove pressure adjusting screw (52) so that it can be reinstalled to original setting.

Check for wear, scoring or scratches on pump and motor pistons and cylinder bores, and on valve port plates (15 and 16) and mating surfaces on pump and motor blocks. Inspect charge pump body, rotor set and wear plate for wear or scoring. Renew any parts that appear questionable. Renew all oil seals and "O" rings.

Before reassembling, be sure that all parts are clean and lubricated with Ford 134 oil. Assemble charge pressure relief valve, high pressure relief valve, feed valves and neutral valves in port plate. Tighten neutral valves to specified torque of 35-39 N·m (26-29 ft.-lbs.).

Install output shaft and bearing in motor housing. Tighten end cover retaining screws to specified torque of 4.5-5.4 N·m (3.5-4.0 ft.-lbs.). It is recommended that a dummy shaft (S—Fig. 166) be used to hold the three pins (13) in place in motor block while cylinder block is being installed on output shaft. Install thrust plate and motor cylinder block assembly into motor housing. When correctly assembled, depth from surface of motor housing to face of cylinder block should be approximately 1.5 mm (0.059 inch) on Models 1120 and 1220 or 2.0 mm (0.078 inch) on Models 1320 and 1520. If not, recheck for proper as-

sembly. Position valve plate on port block, making sure that dowel pin (50—Fig. 164 or 165) is in place in port block. Note that pump valve plate (16) has feathering notches in one end of each oil port passage, while motor valve plate (15) does not. Use lithium base grease to hold valve plate in place. Install motor and housing assembly onto port block and tighten retaining screws to specified torque of 31-38 N·m (23-28 ft.-lbs.). Be sure that output shaft rotates freely.

Assemble input shaft and charge pump assembly in pump housing. Tighten charge pump retaining screws to specified torque of 16-19 N·m (12-14 ft.-lbs.) on all models. Install pump swash plate and covers, aligning reference marks on covers made prior to disassembly. Use a dummy shaft (S—Fig. 166) to hold three pins (18) in position in cylinder block while unit is being installed on input shaft. Install thrust plate

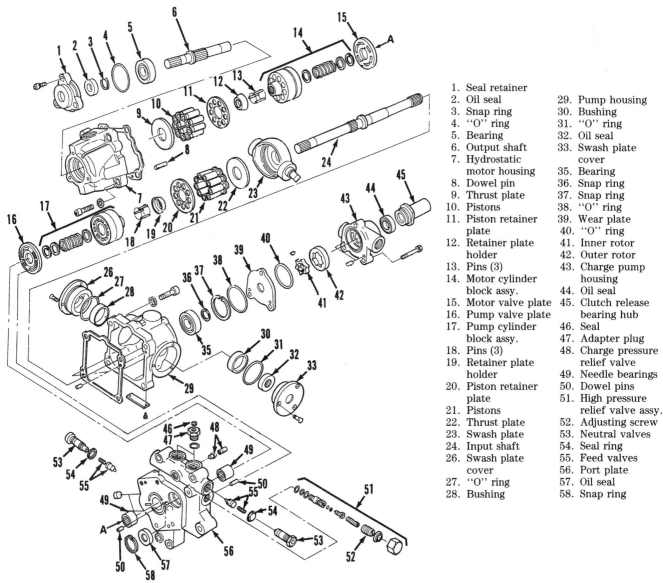

1. Seal retainer	29. Pump housing
2. Oil seal	30. Bushing
3. Snap ring	31. "O" ring
4. "O" ring	32. Oil seal
5. Bearing	33. Swash plate
6. Output shaft	cover
7. Hydrostatic	35. Bearing
motor housing	36. Snap ring
8. Dowel pin	37. Snap ring
9. Thrust plate	38. "O" ring
10. Pistons	39. Wear plate
11. Piston retainer	40. "O" ring
plate	41. Inner rotor
12. Retainer plate	42. Outer rotor
holder	43. Charge pump
13. Pins (3)	housing
14. Motor cylinder	44. Oil seal
block assy.	45. Clutch release
15. Motor valve plate	bearing hub
16. Pump valve plate	46. Seal
17. Pump cylinder	47. Adapter plug
block assy.	48. Charge pressure
18. Pins (3)	relief valve
19. Retainer plate	49. Needle bearings
holder	50. Dowel pins
20. Piston retainer	51. High pressure
plate	relief valve assy.
21. Pistons	52. Adjusting screw
22. Thrust plate	53. Neutral valves
23. Swash plate	54. Seal ring
24. Input shaft	55. Feed valves
26. Swash plate	56. Port plate
cover	57. Oil seal
27. "O" ring	58. Snap ring
28. Bushing	

Fig. 164—Exploded view of hydrostatic transmission assembly used on Models 1120 and 1220.

and pump cylinder block assembly in housing. When correctly assembled, depth from surface of pump housing to face of cylinder block should be approximately 1.5 mm (0.059 inch) on Models 1120 and 1220 or 2.0 mm (0.078 inch) on Models 1320 and 1520. Position valve port plate (16—Fig. 164 or 165) onto port plate block, using small amount of lithium base grease to hold valve plate in place. Be sure that charge pressure relief valve poppet and spring (48) are correctly positioned in port block, then install pump housing assembly. Tighten retaining screws to specified torque of 30.5-37.5 N·m (22.5-27.5 ft.-lbs.). Be sure that input shaft rotates freely.

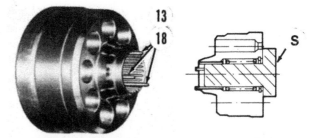

Fig. 166—To facilitate installation of pump and motor, it is recommended that a dummy shaft (S) be used to hold pins (13 and 18) in place while installing pump and motor onto input shaft and output shaft.

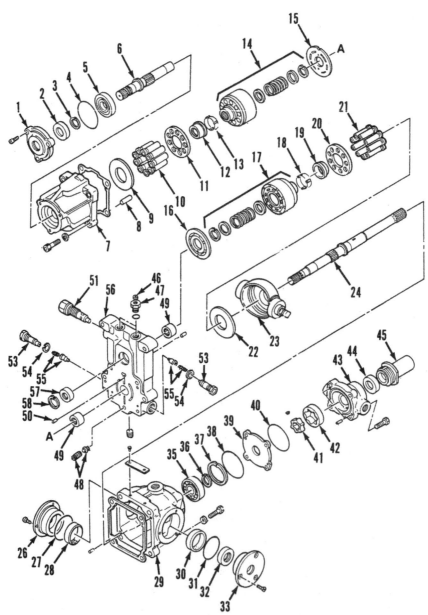

Fig. 165—Exploded view of hydrostatic transmission assembly used on Models 1320 and 1520. Refer to Fig. 164 for legend.

RANGE TRANSMISSION

Models 1120-1220

105. R&R AND OVERHAUL. Drain oil from transmission. Split tractor between transmission extension housing and rear axle center housing as follows: Disconnect wiring from rear lights. Remove ROPS bar, fenders, seat, seat platform, side panels, step plates and transmission cover plate. Remove hydrostatic oil tubes (1 and 7—Fig. 161), hydraulic system pressure tube (3), filter inlet tube (5), suction tube (4) and oil return tube (10). Unbolt and remove hydraulic filter and mounting bracket (2). Remove transmission top cover and hydraulic lift housing. Disconnect brake

rods from both sides. Support rear axle center housing and extension housing with suitable stands and hoist or jack. Remove retaining cap screws and roll rear axle center housing away from front of tractor.

To disassemble, remove transmission drive shaft (49—Fig. 168) assembly. Remove bearing (43) from front of shaft and separate gears (47) and bearings (45) from shaft. Remove retaining cap screw (38) from rear of extension housing and withdraw input shaft (39) from the housing. Remove snap ring (14) and withdraw drive gear (15), thrust washer (16) and idler gear (17) from front of transmission countershaft (21). Remove snap ring (1), drive gear (2) and spacer (3) from front of transmission mainshaft (6). Unbolt and remove bearing retainer housing (19).

Remove snap ring (13) from rear end of mainshaft (6), then slide mainshaft forward and disengage snap

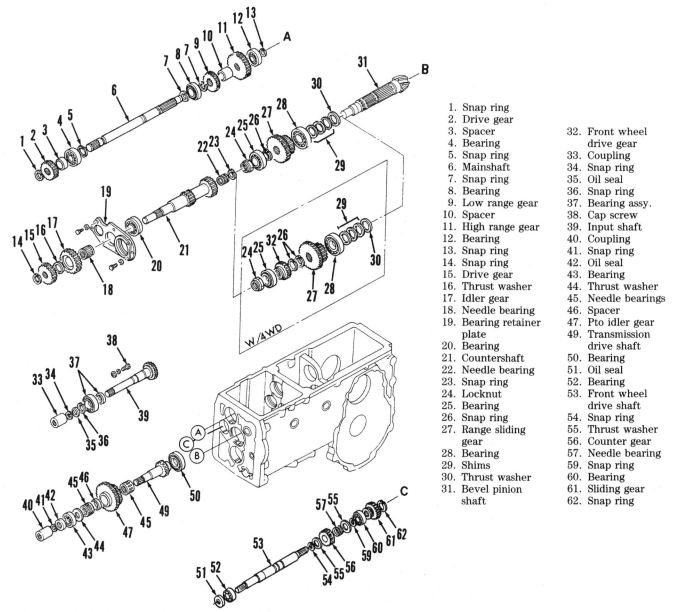

1. Snap ring
2. Drive gear
3. Spacer
4. Bearing
5. Snap ring
6. Mainshaft
7. Snap ring
8. Bearing
9. Low range gear
10. Spacer
11. High range gear
12. Bearing
13. Snap ring
14. Snap ring
15. Drive gear
16. Thrust washer
17. Idler gear
18. Needle bearing
19. Bearing retainer plate
20. Bearing
21. Countershaft
22. Needle bearing
23. Snap ring
24. Locknut
25. Bearing
26. Snap ring
27. Range sliding gear
28. Bearing
29. Shims
30. Thrust washer
31. Bevel pinion shaft
32. Front wheel drive gear
33. Coupling
34. Snap ring
35. Oil seal
36. Snap ring
37. Bearing assy.
38. Cap screw
39. Input shaft
40. Coupling
41. Snap ring
42. Oil seal
43. Bearing
44. Thrust washer
45. Needle bearings
46. Spacer
47. Pto idler gear
49. Transmission drive shaft
50. Bearing
51. Oil seal
52. Bearing
53. Front wheel drive shaft
54. Snap ring
55. Thrust washer
56. Counter gear
57. Needle bearing
59. Snap ring
60. Bearing
61. Sliding gear
62. Snap ring

Fig. 168—Exploded view of range transmission components used on Models 1120 and 1220 equipped with hydrostatic transmission.

ring (7) from groove in mainshaft. Withdraw mainshaft out front of housing while removing gears (9 and 11) and spacer (10). Remove bearings (4, 8 and 12) if necessary.

If equipped with front wheel drive, remove hydraulic system filter and hydrostatic system filter if not already removed. Disconnect front wheel drive shift lever from shift arm, then remove shift cover from side of axle center housing. Remove snap ring (62) from rear end of front wheel drive shaft (53) and remove sliding gear (61). Withdraw drive shaft from front of housing while removing drive gear (56), bearing (57) and thrust washers (55).

Slide countershaft (21) forward and remove bearing (20) from front of shaft, then lift shaft out through top of housing. Remove snap ring (23) and needle bearing (22) from rear of shaft.

Service procedures covering removal, installation and adjustment of bevel pinion shaft (31) are outlined in paragraphs 130 through 133. Refer to paragraphs 145 and 146 for removal of pto front countershaft and mid-mount pto countershaft (if so equipped).

Inspect bearings, gears and shafts for excessive wear or other damage and renew as necessary.

To reassemble range transmission components, reverse the disassembly procedure.

Models 1320-1520

106. R&R AND OVERHAUL. Drain oil from transmission. Split tractor between engine and clutch housing and remove clutch housing and hydrostatic unit from transmission extension housing as outlined in paragraph 103. Remove ROPS roll bar, fenders, seat and seat platform. Remove transmission cover and hydraulic lift housing. Remove snap ring (42—Fig. 170) and drive gear (43) from front of mainshaft. Sup-

Fig. 170—Snap ring (42) and gear (43) must be removed from range transmission mainshaft before separating hydrostatic transmission extension housing from rear axle center housing. Note that one retaining cap screw (C) is located inside the extension housing.

4. Pto input shaft	42. Snap ring
6. Snap ring	43. Mainshaft drive
9. Gear	gear

port rear axle center housing and transmission housing with suitable stands and hoist. Remove retaining cap screws (one cap screw is located inside axle center housing) and separate transmission front housing from rear axle center housing.

To disassemble, remove snap ring (6—Fig. 171) and gear (9) from rear of pto input shaft (4). Remove oil seal (5) and snap ring (7) from front of transmission housing. Drive pto input shaft forward from housing.

Remove oil seal (14—Fig. 171) and snap ring (15) from transmission housing. Drive transmission input shaft (18) and front bearing (16) out front of housing while removing gears (21 and 23), needle bearings (20) and spacer (22) through top of housing.

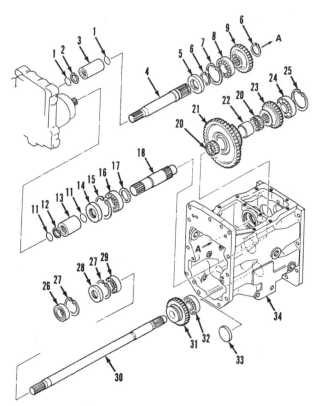

Fig. 171—Pto input shaft (4), transmission input shaft (18) and front wheel drive shaft (30) are located in hydrostatic transmission extension housing (34) on Models 1320 and 1520.

1. "O" rings	
2. Snap ring	20. Needle bearings
3. Coupling	21. Cluster gear
4. Pto input shaft	22. Spacer
5. Oil seal	23. Drive gear
6. Snap ring	24. Bearing
7. Snap ring	25. Snap ring
8. Bearing	26. Bearing
9. Drive gear	27. Snap ring
11. "O" rings	28. Oil seal
12. Snap ring	29. Bearing
13. Coupling	30. Front wheel
14. Oil seal	drive shaft
15. Snap ring	31. Sliding gear
16. Bearing	32. Bearing
17. Thrust washer	33. Plug (w/2WD)
18. Transmission	34. Extension
drive shaft	housing

If equipped with front wheel drive axle, drive roll pin out of front wheel drive shifter lever and remove the lever. Pull shift rail out rear of housing, being careful not to lose detent ball and spring, and remove shift fork. Withdraw shift arm from housing. Remove seal (28—Fig. 171) and snap ring (27) from front of front wheel drive shaft (30). Pull the shaft out front of housing and remove sliding gear (31) and rear bearing (29).

Drive mainshaft (45—Fig. 172) forward out of rear axle center housing while removing rear bearing (41), gears (37, 39 and 40) and spacers (38).

Drive out roll pin and remove range shift lever. Pull shift rod out front of housing, being careful not to lose detent ball and spring, and remove shift fork.

Remove retainer plate and remove shift arm from housing.

Refer to paragraphs 135, 137, 138 and 139 for service procedures outlining removal, installation and adjustment of bevel pinion shaft (46).

To remove pto front countershaft (59—Fig. 172), first remove pto output shaft and lower drive shaft as outlined in paragraph 147. Remove snap ring (64) from front of countershaft, then drive the shaft rearward from axle center housing and while removing thrust washer (60) and front wheel drive counter gear (63).

Inspect all parts for excessive wear or damage and renew as necessary. Renew all "O" rings and oil seals. To reassemble, reverse the disassembly procedure.

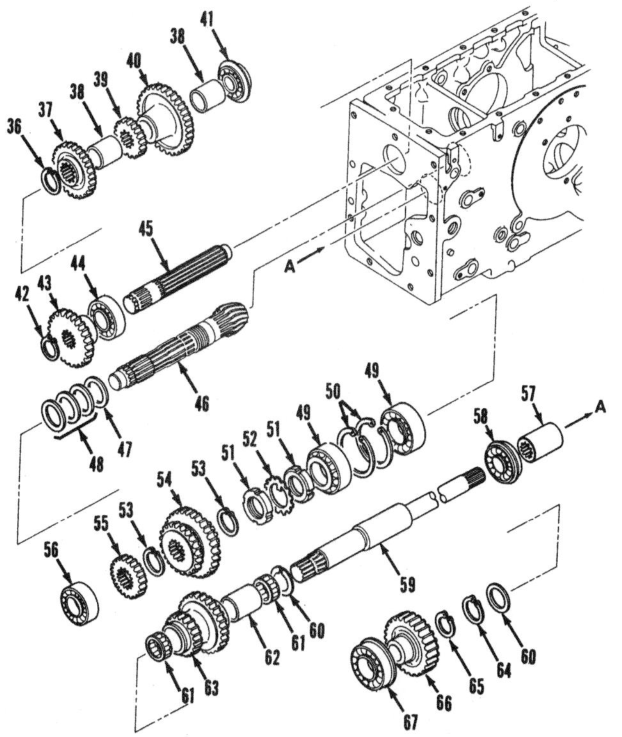

Fig. 172—Range transmission mainshaft (45), bevel pinion shaft (46) and pto countershaft (59) are located in rear axle center housing on Models 1320 and 1520 with hydrostatic transmission.

36. Snap ring	52. Lockwasher
37. Mid-range gear	53. Snap ring
38. Spacer	54. Range sliding gear
39. Low range gear	55. Front wheel drive
40. High range gear	gear
41. Bearing	56. Bearing
42. Snap ring	57. Coupling
43. Drive gear	58. Bearing
44. Bearing	59. Pto countershaft
45. Range mainshaft	60. Thrust washer
46. Bevel pinion	61. Needle bearings
shaft	62. Spacer
47. Thrust washer	63. Counter gear
48. Shims	64. Snap ring
49. Bearing	65. Snap ring
50. Snap rings	66. Pto drive gear
51. Nuts	67. Bearing

NONSYNCHROMESH 12 × 4 GEAR TRANSMISSION

(Models 1720-1920-2120)

The nonsynchromesh transmission that is standard equipment on Models 1720, 1920 and 2120 consists of a main transmission and a range transmission located in separate compartments in transmission housing. The forward compartment contains the main transmission gears which provide three forward speeds and one reverse speed. The rear compartment contains range gears which provide four range speeds for each of the main transmission speeds for a total of 12 forward speeds and four reverse speeds.

LUBRICATION

Models 1720-1920-2120

107. The transmission housing and rear axle center housing serve as a common reservoir for gear lubricant and oil for hydraulic system. The oil should be drained and refilled with Ford 134 oil or equivalent lubricant every 300 hours of operation. Capacity is approximately 27 liters (28.5 U.S. quarts) for Model 1720, 29 liters (30.5 U.S. quarts) for Model 1920 and 33 liters (34.9 U.S. quarts) on Model 2120. Fluid level should be maintained between full mark and lower end of dipstick, which is located in transmission top cover. Oil filler opening is located in rear of hydraulic lift housing.

The hydraulic system filter is located on right side of transmission housing. The spin-on type filter should be renewed every 300 hours of operation also.

REMOVE AND REINSTALL

Models 1720-1920-2120

108. To remove transmission, first drain oil from transmission and rear axle center housings. Remove transmission housing from tractor as follows: Split tractor between engine and clutch housing as outlined in paragraph 89. Remove brake and clutch pedal return springs. Disconnect foot throttle control linkage from pedal. Remove rubber mats and step plates from each side. Disconnect brake control rods and clutch control rod. Unbolt and remove brake cross shaft and pedals as an assembly. Disconnect shift linkage, then unbolt and remove steering column, throttle control linkage and shift linkage as an assembly from transmission housing. Disconnect rear wiring harness. Remove parking brake linkage and switch. Remove ROPS roll bar, fenders, seat, seat support and sheet metal panels. Disconnect range shift links. Remove transmission housing top cover and rear axle housing front cover. Support rear axle center housing and transmission housing with suitable safety stands and hoist or floor jack. Remove cap screws retaining transmission housing to axle center housing (including three cap screws located inside transmission housing). Slide transmission forward off dowel pins.

To reinstall transmission, reverse the removal procedure.

OVERHAUL

Models 1720-1920

109. To disassemble transmission, remove retainer (2—Fig. 174) from front of transmission housing. If equipped with double disc clutch, remove pto input shaft (27—Fig. 175). Drive the countershaft (18) rearward, removing front bearing (11), gears (13, 15 and 16) and spacer (14) as shaft is removed. Remove bearing retaining snap rings (12) from center webs of housing, then drive transmission input shaft (20) rearward from housing while removing gears (3, 5, 6, and 9), spacer (4) and collar assembly (7 and 8).

If equipped with single disc clutch, withdraw input shaft (10—Fig. 175) from front of housing while removing gears (3, 5, 6 and 9), spacer (4) and collar assembly (7 and 8). Drive countershaft (18) rearward, removing front bearing (11), gears (13, 15 and 16) and spacer (14) as shaft is removed.

Drive roll pin out of upper shift fork (9—Fig. 174) and rail (7). Remove detent pin (3), spring (4) and ball (5). Place lower shift rail (12) in neutral position, then slide top rail forward from housing and remove shift fork. Drive roll pin out of lower shift rail (12) and shift boss (10), rotate shift rail 90 degrees and slide rail out

front of housing. Remove lower detent ball (5) and spring (4) from housing. Remove safety wire from shift arm (15), then drive out roll pins and remove shift arm and shaft (14).

Remove range shift rail retaining bolt (18—Fig. 174) and shift arm retaining plate (17) from transmission housing. Rotate shift rail (19) 180 degrees and slide rail rearward from housing. Remove range shift forks (23 and 24).

Drive rear mainshaft (45—Fig. 176) and bearing (44) rearward from housing while removing range sliding gears (41 and 42). Remove bearing retaining snap ring (38) from housing, then drive front mainshaft (39) rearward to provide clearance for removal of sliding gears (31, 32, 34 and 35) and spacer (33) from front of shaft. Remove reverse idler shaft retaining bolt (46) and slide reverse idler gear (51) and shaft (48)

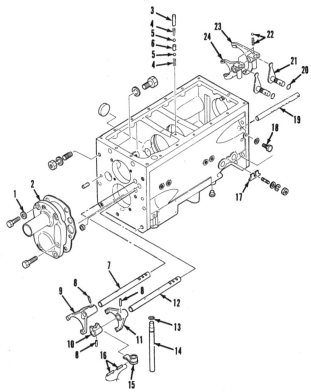

Fig. 174—Exploded view of main gearshift forks and rails used on Models 1720 and 1920 with 12 × 4 gear transmission.

1. Seal washer	14. Shift shaft
2. Bearing retainer	15. Shift arm
3. Detent pin	16. Pins
4. Spring	17. Retainer plate
5. Detent ball	18. Retainer bolt
6. Interlock pin	19. Shift rail
7. Shift rail	20. "O" ring
8. Pin	21. Shift arm
9. Shift fork, 2nd-rev.	22. Detent ball & spring
10. Shift boss	23. Range shift fork, 1st-2nd
11. Shift fork, 1st-3rd	24. Range shift fork, 3rd-4th
12. Shift rail	
13. "O" ring	

rearward to provide clearance for mainshaft rear bearing, then slide front mainshaft and bearing forward from transmission housing. Drive roll pin (49) out of reverse idler shaft. Slide shaft out front of housing and remove thrust washers (50 and 53), idler gear (51) and needle bearings (52).

Inspect all parts for excessive wear or other damage and renew as necessary. Renew all gaskets and seals. Lubricate all parts with clean oil prior to reassembly. Reassemble transmission in the following order: Install reverse idler gear (51—Fig. 176) and shaft (48), but do not install idler shaft retaining bolt at this time. Push idler gear and shaft rearward to provide clearance for installation of mainshaft. Install rear bearing (37) and snap rings (36) on front mainshaft (39), then install mainshaft in housing from the front so that rear bearing is beyond the housing web.

Assemble sliding gears and spacer on mainshaft, then slide mainshaft forward until rear snap ring (38) can be installed in housing web. Align idler shaft counterbore with locking bolt and install bolt with new seal washer. Position range sliding gears in transmission housing, then install rear mainshaft (45) from the rear. Install range gear shift forks and rail and main gear shift forks and rails.

If equipped with double disc clutch, insert input shaft (20—Fig. 175) from the rear through bearings, gears (3, 5, 6 and 9), spacer (4) and adjustable collar (7 and 8). Then install countershaft (18), bearings and gears (13, 15 and 16).

> **NOTE: If new input shaft components are being installed or if adjustable collar and nut assembly (7 and 8) has been separated, renew collar and nut assembly. When using a new collar, adjust collar and nut to provide free play of gears while assembling input shaft.**

If equipped with single disc clutch, install countershaft (18—Fig. 175), bearings and gears first, then po-

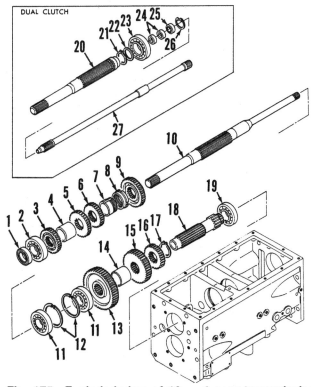

Fig. 175—Exploded view of 12 × 4 gear transmission upper shafts and related components used on Models 1720 and 1920.

1. Oil seal	14. Spacer
2. Bearing	15. Range gear, 3rd
3. Reverse gear	16. Range gear, 2nd
4. Spacer	17. Snap ring
5. Second gear	18. Countershaft
6. First gear	19. Bearing
7. Adjustable collar	20. Transmission input
8. Nut	shaft
9. Third gear	21. Snap ring
10. Input shaft	22. Spacer
(w/single disc	23. Bearing
clutch)	24. Oil seals
11. Bearings	25. Bearing
12. Snap rings	26. Snap ring
13. Range gear, 4th	27. Pto input shaft

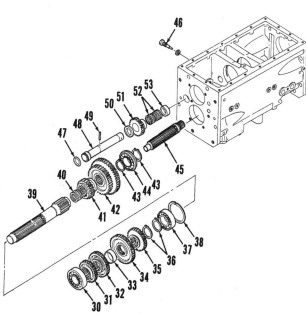

Fig. 176—Exploded view of 12 × 4 gear transmission lower shafts and reverse idler used on Models 1720 and 1920.

30. Bearing		
31. Sliding gear, rev.		
32. Sliding gear, 2nd	43. Snap ring	
33. Spacer	44. Bearing	
34. Sliding gear, 1st	45. Snap ring	
35. Sliding gear, 3rd	46. Rear mainshaft	
36. Snap rings	47. "O" ring	
37. Bearing	48. Idler shaft	
38. Snap ring	49. Pin	
39. Front mainshaft	50. Thrust washer	
40. Needle bearing	51. Reverse idler	
41. Sliding gear, 3rd-	gear	
4th range	52. Needle bearings	
42. Sliding gear, 1st-	53. Thrust washer	
2nd range	54. Retaining bolt	

sition gears (3, 5, 6 and 9), spacer (4) and collar assembly (7 and 8) in housing. Install input shaft (10) from the front through gears and collars.

On all models, install new oil seal in front retainer (2—Fig. 174). Install retainer using new seal washers (1) on retaining cap screws, then adjust length of collar assembly (7 and 8—Fig. 175) to obtain zero free play and zero preload of gears on input shaft. When properly adjusted, the collar should rotate with medium hand pressure. Stake flange of collar nut into groove in collar to lock the assembly.

Model 2120

110. To disassemble transmission, remove retainer (1—Fig. 177) and clutch release bearing from front of housing. Drive pto input shaft (1—Fig. 178) rearward from housing. Drive countershaft (20) and rear bearing (21) rearward from housing and remove gears (15, 18 and 19) and adjustable collar assembly (16 and 17). Disengage snap ring (4) from groove in input shaft (10) and slide snap ring forward on shaft. Drive

input shaft rearward while removing snap ring (4), gears (5, 7, 8 and 9) and spacers (6) from front of shaft.

Remove range shift rail retaining bolt (24—Fig. 177) and shift arm guide (21) from left side of transmission housing. Cover shift forks (14 and 15) with shop cloth to contain detent balls. Rotate shift rail (18) 90 degrees, then withdraw rail from rear of housing and remove shift forks. Drive roll pins (5) out of main transmission shift forks (7 and 10) and shifter boss (6 and 9). Remove detent pins (2), springs (3) and balls (4) from top of housing, then withdraw shift rails (8 and 11) from front of housing.

Drive roll pin (51—Fig. 179) out of reverse idler shaft (50). Remove idler shaft retaining bolt (48), then slide idler shaft out front of transmission housing and remove idler gear (52), bearings (53) and thrust washer (54).

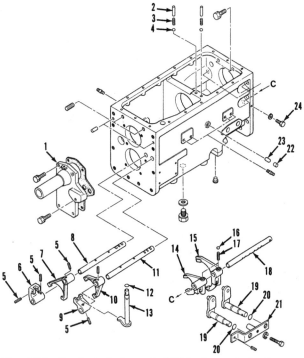

Fig. 177—Exploded view of shift forks and shafts used on 12 × 4 gear transmission used on Model 2120.

1. Bearing retainer	14. Range shift fork
2. Detent pin	15. Range shift fork
3. Spring	16. Detent ball
4. Detent ball	17. Spring
5. Pins	18. Shift rail
6. Shift boss	19. Shift arms
7. Shift fork	20. "O" rings
8. Shift rail	21. Shift guide
9. Shift boss	22. Plug
10. Shift fork	23. Shift interlock
11. Shift rail	pin
12. "O" ring	24. Shift rail
13. Shift shaft	retaining bolt

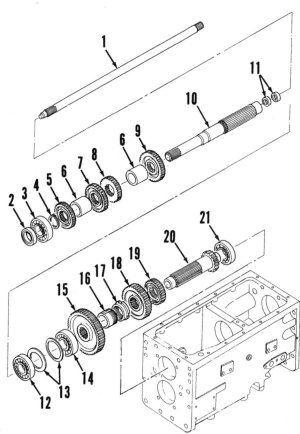

Fig. 178—Exploded view of upper shafts and related components used on Model 2120 12 × 4 gear transmission.

1. Pto input shaft	12. Bearing
2. Oil seal	13. Spacers
3. Bearing	14. Bearing
4. Snap ring	15. Drive gear
5. Reverse gear	16. Adjustable collar
6. Spacer	17. Nut
7. Second gear	18. Range gear, 3rd
8. First gear	19. Range gear, 2nd
9. Third gear	20. Countershaft &
10. Transmission	range gear, 1st
input shaft	21. Bearing
11. Oil seals	

Remove snap ring (44—Fig. 179) from rear of housing. Drive rear mainshaft (47) out rear of housing while removing sliding gears (42 and 43) off front of shaft. Remove range shift arms (19—Fig. 177) from inside housing if necessary. Remove snap ring (39—Fig. 179) from housing center web. Drive front mainshaft (40) out rear of housing and remove counter gear (29), sliding gears (32, 33, 35 and 36), spacers and thrust washers through top opening.

Inspect all parts for excessive wear or other damage and renew as necessary. Renew all seals and gaskets. Lubricate parts with clean oil during assembly.

To reassemble, insert front mainshaft (40—Fig. 179) from the rear while installing sliding gears, spacers, thrust washers and counter gear onto shaft. Be sure that grooved face of thrust washers (27) is positioned toward counter gear (29). If removed, install range shift arms (19—Fig. 177) in housing prior to install-

ing rear mainshaft (47—Fig. 179). Insert rear mainshaft from the rear through the sliding gears and into needle bearing (41) in end of front mainshaft. Install reverse idler gear (52) and shaft (50), making sure that grooved side of thrust washer (54) faces the idler gear. Install idler shaft retaining bolt (48) using a new seal washer, then drive roll pin (51) into idler shaft. Install main transmission shift forks and rails and range gear shift forks and rail (Fig. 177).

Install bearing (12—Fig. 178) on rear of transmission input shaft (10), then insert shaft into housing from the rear while assembling gears (5, 7, 8 and 9), spacers (6) and snap ring (4) onto shaft. Install front bearing (3) and retainer housing (1—Fig. 177). Position shims (13—Fig. 178) and bearing (14) in housing center web. Insert countershaft (20) from the rear while assembling gears (15, 18 and 19) and adjustable collar assembly (16 and 17) onto shaft.

NOTE: If any of the input shaft (10—Fig. 178) or countershaft (20) components are renewed, adjustable collar assembly (16 and 17) should also be renewed.

Make sure that countershaft (20—Fig. 178) is seated properly in front bearing (14). Install pto input shaft coupling bearing (1—Fig. 180) in rear of housing. Use

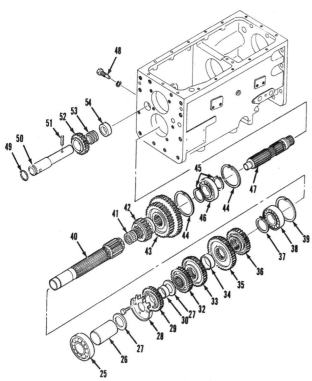

Fig. 179—Exploded view of Model 2120 12 × 4 gear transmission lower shafts and related components.

25. Bearing	42. Range sliding gear, 3rd-4th
26. Spacer	
27. Thrust washer	43. Range sliding gear, 1st-2nd
28. Oil slinger	
29. Counter gear	44. Snap ring
30. Bushing	45. Snap rings
32. Sliding gear, rev.	46. Bearing
33. Sliding gear, 2nd	47. Rear mainshaft
34. Spacer	48. Retaining bolt
35. Sliding gear, 1st	49. "O" ring
36. Sliding gear, 3rd	50. Idler shaft
37. Snap ring	51. Pin
38. Bearing	52. Reverse idler gear
39. Snap ring	
40. Front mainshaft	53. Needle bearing
41. Needle bearing	54. Thrust washer

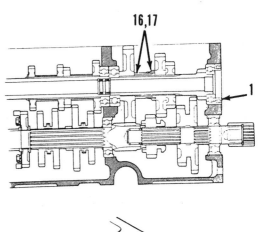

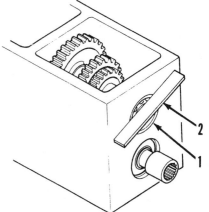

Fig. 180—Shorten or lengthen adjustable collar and nut (16 and 17) to position pto input shaft rear bearing (1) flush with rear surface of transmission housing.

a straightedge to make sure that coupling bearing does not protrude beyond rear surface of transmission housing. If necessary, shorten collar assembly (16 and 17) to eliminate any protrusion. When correctly adjusted, stake flange of collar nut into groove in collar to lock the assembly.

Lubricate pto input shaft (1—Fig. 178) with oil, then insert shaft from the front through transmission input shaft and countershaft. Rotate pto input shaft as it is being installed to prevent damage to oil seals (2 and 11). Install drive coupling on rear of pto input shaft.

SYNCHROMESH 12 × 12 SHUTTLE TRANSMISSION

(Models 1720-1920)

An optional synchromesh shuttle transmission is available on Models 1720 and 1920. The transmission gears are contained in three compartments in transmission housing (Fig. 181). The forward compartment contains forward and reverse shuttle gears. The shuttle gears are synchronized and controlled by a lever on steering column. The center compartment contains main transmission gears consisting of four speed ratios. The main transmission gears are synchronized and controlled by a lever on steering column. The rear compartment contains the three speed range gears. The range gears are not synchronized and are controlled by a lever on left side of transmission housing. The shuttle transmission provides 12 speeds in forward and 12 speeds in reverse.

LUBRICATION

Models 1720-1920

112. The transmission housing and rear axle center housing serve as a common reservoir for gear lubricant and oil for hydraulic system. The oil should be drained and refilled with Ford 134 oil or equivalent lubricant every 300 hours of operation. Fluid lev-

Fig. 181—Shuttle shift 12 × 12 transmission used on Models 1720 and 1920 is contained in three separate compartments in transmission housing.

1. Shuttle
   transmission gears
2. Main transmission
   gears
3. Range
   transmission gears

el should be maintained between full mark and lower end of dipstick, which is located in transmission top cover. Oil filler opening is located in rear of hydraulic lift housing.

The hydraulic system filter is located on right side of transmission housing. The spin-on type filter should be renewed every 300 hours of operation also.

REMOVE AND REINSTALL

Models 1720-1920

113. To remove transmission, first drain oil from transmission and rear axle center housings. Remove transmission housing from tractor as follows: Split tractor between engine and clutch housing as outlined in paragraph 87. Remove brake and clutch pedal return springs. Disconnect foot throttle control linkage from pedal. Remove rubber mats and step plates from each side. Disconnect brake control rods and clutch control rod. Unbolt and remove brake cross shaft and pedals as an assembly. Disconnect shift linkage, then unbolt and remove steering column, throttle control linkage and shift linkage as an assembly from transmission housing. Disconnect rear wiring harness. Remove parking brake linkage and switch. Remove ROPS roll bar, fenders, seat, seat support and sheet metal panels. Disconnect range shift links. Remove transmission housing top cover and rear axle housing front cover. Support rear axle center housing and transmission housing with suitable safety stands and hoist or floor jack. Remove cap screws retaining transmission housing to axle center housing (including three cap screws located inside transmission housing). Slide transmission forward off dowel pins.

To reinstall transmission, reverse the removal procedure.

OVERHAUL

Models 1720-1920

114. To disassemble shuttle transmission, remove seal retainer (1—Fig. 182) from front of housing. Drive roll pins (5) out of upper shift fork (8) and boss (7).

Remove detent spring (2) and ball (3) from housing bore, then slide upper shift rail (9) rearward from case and remove shift fork and boss. Be careful not to lose detent ball as shift rail is removed from shift fork. Remove interlock pin (4) from housing bore. Drive roll pins (5) out of lower shift fork (11) and boss (12). Rotate lower shift rail (10) 90 degrees, then slide rail rearward from housing and remove shift fork and boss.

Withdraw pto drive shaft (31—Fig. 183) from rear of transmission housing. Drive rear bearing (32) rearward from housing. Disengage snap ring (29) from groove in mainshaft (30), then drive mainshaft rearward until center bearing (24) is out of housing web. Withdraw shaft from rear of housing while removing range gears (25 and 28), collar assembly (26 and 27), center bearing (24), spacers and synchronizer assemblies (12 through 23), front bearing (10) and snap ring (11).

NOTE: Prior to disassembling synchronizer assemblies, place matching marks on synchronizer hub and sliding sleeve so that they can be reassembled in original positions.

Remove shift rail retaining bolt (33—Fig. 182), rotate range shift rail (28) 90 degrees and slide shift rail from rear of housing. Remove range shift fork (25). Slide snap ring (54—Fig. 184) forward on range gear countershaft (59), then withdraw countershaft rearward from housing. Lift out sliding gear (53), drive gear (55), spacer (56) and snap ring (54).

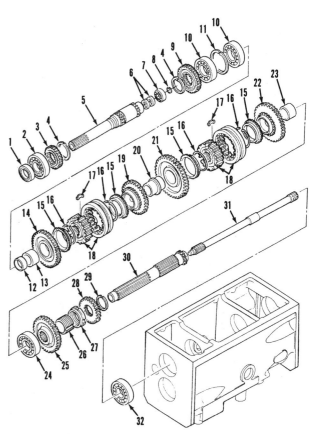

Fig. 183—Exploded view of 12 × 12 shuttle shift transmission upper shafts and related components used on Models 1720 and 1920.

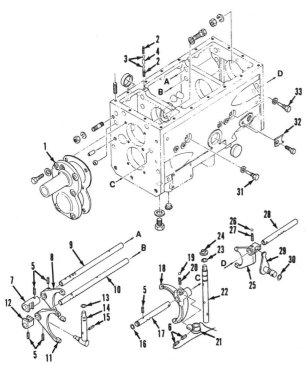

Fig. 182—Exploded view of shift components used on 12 × 12 shuttle shift transmission on Models 1720 and 1920.

1. Bearing retainer
2. Springs
3. Detent balls
4. Shift interlock pin
5. Pins
6. Pins
7. Shift boss
8. Shift fork, 1st-2nd
9. Shift rail
10. Shift rail
11. Shift fork, 3rd-4th
12. Shift boss
13. "O" ring
14. Shift shaft
15. Pin
16. "O" ring
17. Shift rail
18. Shuttle shift fork
19. Detent ball
20. Spring
21. Shift arm
22. Shift shaft
23. "O" ring
24. "C" ring
25. Range shift fork
26. Detent ball
27. Spring
28. Shift rail
29. Shift arm
30. "O" ring

1. Oil seal
2. Bearing
3. Reverse gear
4. Snap ring
5. Input shaft
6. Oil seals
7. Bearing
8. Snap ring
9. Forward gear
10. Bearings
11. Snap ring
12. Spacer
13. Collar
14. Synchronizer gear, 3rd
15. Synchronizer ring
16. Spring
17. Key
18. Synchronizer hub & sliding collar
19. Synchronizer gear, 4th
20. Collar
21. Synchronizer gear, 1st
22. Synchronizer gear, 2nd
23. Collar
24. Bearing
25. High range gear
26. Adjustable collar
27. Nut
28. Mid-range gear
29. Snap ring
30. Mainshaft
31. Pto input shaft
32. Bearing

Pull input shaft (5—Fig. 183) forward and remove bearing (10) from rear of shaft. Slide gear (9), snap rings (4) and gear (3) off rear of input shaft as shaft is removed from housing.

Drive roll pin (63—Fig. 184) out of idler gear shaft (62). Remove idler shaft retaining bolt (60), then drive the shaft out front of housing and remove thrust washer (66), reverse idler gear (65) and needle bearing (64).

Remove rubber plug from bore in front of transmission housing. Remove locking wire from shuttle shift arm (21—Fig. 182). Insert a punch through access hole in front of housing and drive roll pins (6) out of shift arm and shaft (22). Withdraw shift shaft and arm

from housing. Drive roll pin out of shift rail (17) and remove retaining bolt (31). Turn shift rail 90 degrees, then slide shift rail out of shift fork (18) being careful not to lose detent ball (19).

Remove locknut (36—Fig. 184) from front of main countershaft (49). Remove snap ring (52) retaining rear bearing (51) in housing web. Drive countershaft rearward, removing front bearing (38), synchronizer assembly (39 through 46), center bearing (47) and drive gear set (48) as shaft is withdrawn from housing.

Inspect all parts for excessive wear or other damage and renew as necessary. Position synchronizer ring (1—Fig. 185) on its mating synchronizer gear (2) and measure clearance (C) using a feeler gage (3) as shown. Renew synchronizer ring and gear if clearance is less than 0.8 mm (0.031 inch). Measure side clearance (D—Fig. 186) between shift forks (1) and

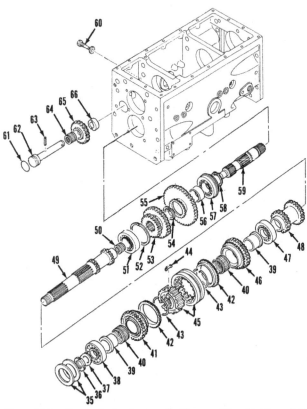

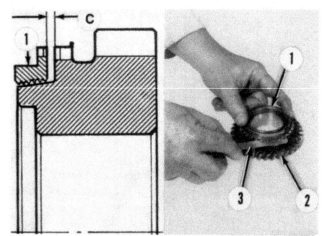

Fig. 185—To check synchronizer ring (1) and gear (2) for wear, measure clearance (C) between ring and gear. Minimum allowable clearance is 0.5 mm (0.020 inch).

Fig. 184—Exploded view of 12 × 12 shuttle shift transmission lower shafts and related components.

35. Shims	
36. Locknut	53. Range sliding
37. Lockwasher	gear, high-
38. Bearing	medium
39. Collar	54. Snap ring
40. Needle bearing	55. Range sliding
41. Counter gear	gear, low
42. Synchronizer ring	56. Bushing
43. Spring	57. Bearing
44. Key	58. Snap ring
45. Synchronizer hub	59. Range
& sliding collar	countershaft
46. Counter gear	60. Retaining bolt
47. Bearing	61. "O" ring
48. Counter gear	62. Idler shaft
49. Main	63. Pin
countershaft	64. Needle bearing
50. Needle bearing	65. Reverse idler
51. Bearing	gear
52. Snap ring	66. Thrust washer

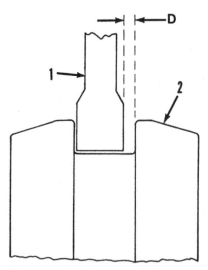

Fig. 186—Measure side clearance (D) between shift forks (1) and synchronizer sliding collars (2). Maximum allowable clearance is 1.0 mm (0.039 inch).

their mating sliding collars (2). Renew fork and/or collar if clearance exceeds 1.0 mm (0.039 inch). Renew all oil seals and "O" rings. Lubricate all parts with clean oil prior to assembly.

Assemble synchronizers and gears, aligning matching marks on synchronizer hubs and sleeves made prior to disassembly. Position shuttle shift synchronizer assembly, gears (41 and 46—Fig. 184), center bearing (47) and gear set (48) in the housing. Insert main countershaft (49) from the rear through gears, bearing and synchronizer assembly. Install front and rear bearings (38 and 51) and rear snap ring (52). Tighten locknut (36) securely and bend tab of lockwasher (37) to prevent loosening. With countershaft positioned fully rearward, measure depth from surface of transmission housing to countershaft front bearing (38). Select shims (35) equal to measured depth to provide zero end play and zero preload of main countershaft assembly. Reinstall shuttle shift rail (17—Fig. 182), shift fork (18), shift shaft (22) and shift arm (21). Be sure to install new lock wire through shift arm roll pins (6) to prevent pins from working out of arm.

Install reverse idler gear (65—Fig. 184) and shaft (62), making sure that grooved side of thrust washer (66) is facing toward idler gear.

Insert input shaft (5—Fig. 183) from the front while assembling gears (3 and 9), snap rings (4) and bearings on shaft. Install front retainer (1—Fig. 182) with correct thickness shims (35—Fig. 184) as determined above. Insert range countershaft (59) from the rear while assembling spacer, gears (53 and 55) and snap ring (54) on shaft. Install range shift fork and rail if removed. Position snap ring (11—Fig. 183) in housing front web.

NOTE: Snap ring (11—Fig. 183) is used as a spacer between bearings (10) and is not positioned in a groove.

Position rear bearing (10—Fig. 183) in housing web. Place spacers, gears, synchronizer assemblies (12 through 23) and center bearing (24) in housing.

NOTE: If any of mainshaft components are renewed, a new adjustable collar assembly (26 and 27) should be installed.

Position snap ring (29—Fig. 187) on mainshaft (30), then insert mainshaft from the rear through gears (25 and 28), adjustable collar (26), center bearing and synchronizer assembly. Note that center bearing (24) must not enter the housing web until mainshaft rear gear (28) is clear of rear range gear (55). Install snap ring (29) into groove in shaft. Install rear bearing (32—Fig. 183). Insert pto drive shaft (31) from the rear.

After transmission housing is reattached to rear axle center housing, adjust mainshaft end play as fol-

lows: Turn adjustable collar nut (27—Fig. 183) as necessary until end clearance is zero and there is zero preload on bearings. Stake flange of nut into groove in collar to lock the assembly.

Reinstall main shift forks, bosses and shift rails.

Fig. 187—When reinstalling mainshaft (30), range gear (28) must be clear of range counter gear (55) on lower shaft before center bearing (24) enters case web.

18. Synchronizer sliding collars	28. Mid-range gear
24. Bearing	29. Snap ring
25. High range gear	30. Mainshaft
26. Adjustable collar	55. Range counter gear

SYNCHROMESH 12 × 12 HYDRAULIC SHIFT SHUTTLE TRANSMISSION

(Model 2120)

An optional synchromesh, hydraulic shift shuttle transmission is available on Model 2120. The forward-reverse shuttle gears (4 and 7—Fig. 190) are located in shuttle clutch housing and are shifted by two hydraulically actuated clutch packs (5 and 6). The four-speed synchronized main transmission gears are combined with three-speed nonsynchronized range gears to provide 12 forward speeds and 12 reverse speeds. The main transmission gears and range gears are located in front and rear compartments in the transmission housing.

LUBRICATION

Model 2120

115. The transmission housing and rear axle center housing serve as a common reservoir for gear lubricant and oil for hydraulic system. The oil should be drained and refilled with Ford 134 or equivalent lubricant every 300 hours of operation. Fluid level should be maintained between full mark and lower end of dipstick, which is located in transmission top cover. Oil filler opening is located in rear of hydraulic lift housing.

The hydraulic system filter is located on right side of transmission housing. The spin-on type filter should be renewed every 300 hours of operation also.

TROUBLE-SHOOTING

Model 2120

116. The following are some problems which may occur when operating hydraulic shift shuttle transmission and their possible causes. Refer to Fig. 193 for identification of control valve components.

1. Shuttle clutch slips or will not engage.
   a. Oil filter plugged.
   b. Relief valve (4) worn, stuck or misadjusted.
   c. Hydraulic pump faulty.
   d. Filter screen (7) or orifice (8) plugged.
   e. Pressure regulating valve (37) not seating properly.
   f. Pressure regulating piston (31) stuck.
   g. Shuttle clutch piston seal rings damaged.
   h. Shuttle clutch discs and plates worn.
2. Transmission main synchronized gears hard to shift.
   a. Lube cut-off valve (27) control cable adjusted incorrectly.

RELIEF VALVE PRESSURE TEST

Model 2120

117. To check hydraulic shuttle relief valve pressure setting, first operate tractor until hydraulic oil is at normal operating temperature. Remove instrument panel and steering column shroud assembly. Disconnect pump pressure tube (1—Fig. 191) from control valve and install a 3000 kPa (500 psi) pressure gage (4) in the pressure line using a tee fitting (2). Position main transmission shift lever in neutral position. Start engine and run at high idle speed.

Fig. 191—To check shuttle shift hydraulic pressure, connect pressure gage (4) to pressure line (1) using a "T" fitting (2).

1. Shuttle shift control valve
2. Transmission drive shaft
3. Clutch release bearing
4. Reverse counter gear
5. Reverse clutch pack
6. Forward clutch pack
7. Forward counter gear
8. Counter gear, 3rd-4th
9. Countershaft & gear, 1st-2nd
10. Range sliding gear, high-middle
11. Range sliding gear, low
12. Rear main shaft
13. Reverse drive gear
14. Forward drive gear
15. Main gear, 3rd
16. Main gear, 4th
17. Main gear, 1st
18. Main gear, 2nd
19. Range gear, high
20. Range gear, middle
21. Countershaft & low range gear

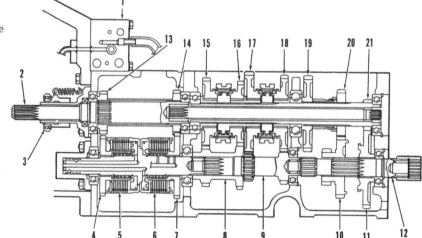

Fig. 190—Cross-sectional drawing of hydraulic shift shuttle transmission available on Model 2120.

Shift shuttle to forward and reverse positions while observing pressure gage reading. Relief valve pressure should be 1670-1765 kPa (242-256 psi) in both directions.

To adjust relief valve pressure, loosen locknut and turn relief valve adjusting screw (1—Fig. 193) clockwise to increase pressure. One full turn of screw will change pressure setting approximately 345 kPa (50 psi).

SHUTTLE CONTROL VALVE

Model 2120

118. R&R AND OVERHAUL. To remove control valve, first remove steering column shroud. Disconnect power steering lines from steering control valve. Disconnect pressure tube and oil return hoses (7—Fig. 192) from shuttle control valve (5). Cap all openings to prevent entry of dirt. Disconnect shuttle shift rod at the universal joint (3). Disconnect shift link (6) from control valve spool. Unbolt and remove shifter bracket (4). Disconnect clutch cable from lube cut-off valve spool. Remove valve mounting cap screws and remove valve from tractor.

Fig. 192—Hydraulic shuttle shift control valve (5) is mounted on top of clutch housing.

1. Shift lever
2. Power steering valve
3. Universal joint
4. Shift rod bracket
5. Shift control valve
6. Shift link
7. Oil return tubes
8. Steering column
9. Main gearshift linkage

To disassemble valve, remove retainer plate (10—Fig. 193), spool cap (22) and detent assembly. Pull control valve spool (12) out of valve body. Remove seals (11 and 13). Remove snap rings (28 and 40). Install a 5 mm bolt in end of pressure regulating valve seat (39) and pull seat out of valve body. Remove valve (37) and springs (36), then push valve piston (31) and plug (29) from valve body. Unscrew bushing (24) from valve body and pull lube cut-off valve spool (27) out of valve body. Remove plug (32), spring (34) and check valve (35). Remove plug (5), snap ring (6) and filter screen (7), then use 2 mm Allen wrench to remove orifice (8). Unscrew relief valve adjusting screw (1) and remove spring (3) and relief poppet (4).

Inspect all components for scratches, scoring or excessive wear and renew as necessary. Control valve assembly must be renewed if control valve spool (12) or valve body (9) are damaged. Pressure regulating valve (37) and seat (39) should also be renewed as a set. Renew all "O" rings and seals. Lubricate components with clean oil prior to assembly.

To reassemble, reverse the disassembly procedure. Apply Loctite 271 to threads of filter screen plug (5).

Install control valve on tractor and adjust lube cut-off valve as outlined in paragraph 119. Adjust relief valve opening pressure as outlined in paragraph 117.

119. LUBE CUT-OFF VALVE ADJUSTMENT. Slowly depress clutch pedal until all slack is removed from cut-off valve cable and valve spool is about to be moved. Clutch pedal should be approximately 25 mm (1 inch) beyond midpoint of clutch pedal travel distance. Measure dimension (D—Fig. 194) between cable clevis (4) and cable bracket (5). Fully depress clutch pedal, then measure dimension (D) again. The difference between the two measurements should be 10-11 mm (3/8-7/16 inch). To adjust cable, loosen locknut and rotate turnbuckle (2) to lengthen or shorten cable as required.

HYDRAULIC PUMP

Model 2120

120. REMOVE AND REINSTALL. The hydraulic pump for hydraulic shuttle transmission is rated at 10 liters/minute (2.6 U.S. gallons/minute) at 2500 engine rpm.

To remove hydraulic pump (3—Fig. 195), raise engine hood and remove left side screen and lower panel. Disconnect the two power steering tubes (2) from side frame. Disconnect inlet tube flange (1) and pressure tube (4) from pump. Remove four retaining nuts and withdraw pump from engine timing gear cover. If necessary, remove adapter plate and pump drive gear with bearings from timing gear cover.

Refer to Fig. 196 for an exploded view of pump assembly. Prior to disassembling pump, scribe match-

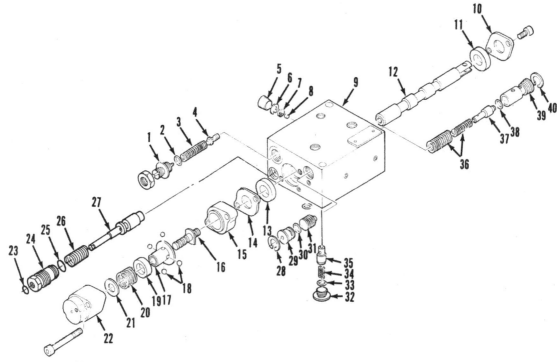

Fig. 193—Exploded view of shuttle shift hydraulic control valve assembly.

1. Adjusting screw	9. Control valve body	17. Detent guide	33. "O" ring
2. "O" ring	10. Plate	18. Balls	34. Spring
3. Spring	11. Seal	19. Bushing	35. Check valve
4. Relief valve plunger	12. Valve spool	20. Spring	36. Springs
5. Plug	13. Seal	21. Shim	37. Pressure regulating valve
6. Snap ring	14. Plate	22. End cap	38. "O" ring
7. Screen	15. Spacer	23. "O" ring	39. Valve seat
8. Orifice	16. Detent plunger	24. Plug	40. Snap ring
		25. "O" ring	
		26. Spring	
		27. Lube cut-off valve spool	
		28. Snap ring	
		29. Plug	
		30. "O" ring	
		31. Piston	
		32. Plug	

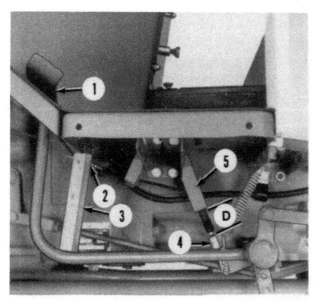

Fig. 194—Hydraulic shuttle lube cut-off valve is actuated by the clutch pedal. Refer to text for adjustment procedure.

1. Clutch pedal
2. Control cable turnbuckle
3. Wood block
4. Clevis
5. Cable bracket

Fig. 195—View of hydraulic shuttle hydraulic pump on Model 2120.

1. Inlet tube
2. Power steering tubes
3. Shuttle shift pump
4. Pressure tube
5. Power steering pump

ing marks on pump body and end cover to ensure correct alignment when reassembling. The only parts serviced on hydraulic pump are shaft seal and internal "O" ring seals. If pump components are excessively worn or damaged, renew pump assembly.

If bearings (6—Fig. 197) were removed from pump drive gear (7), install bearings on gear with sealed side of bearings facing away from the gear. Install pump adapter plate (2) using new "O" rings (5). Be sure that grooves (G) in adapter plate are located at the top. Install pump and reconnect hydraulic lines.

REMOVE AND REINSTALL TRANSMISSION

Model 2120

121. Drain oil from transmission and rear axle center housing. Split tractor between engine and clutch

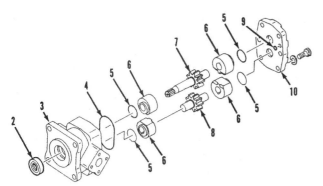

Fig. 196—Exploded view of shuttle shift hydraulic pump assembly.

2. Oil seal
3. Pump housing
4. Seal ring
5. "O" rings
6. Bearings
7. Pump drive gear
8. Idler gear
9. "O" ring
10. End cover

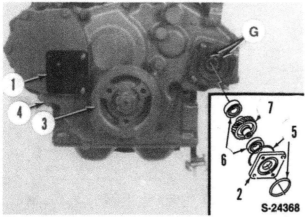

S-24368

Fig. 197—When reinstalling hydraulic pump adapter plate (2), make sure that grooves (G) are positioned at the top.

housing as outlined in paragraph 87. Disconnect shift linkage from transmission. Unbolt and remove steering column, shift linkage and throttle linkage as an assembly. Disconnect foot throttle linkage, brake and clutch pedal return springs, brake rods, clutch rod and lube cut-off valve cable. Unbolt and remove brake cross shaft and pedals as an assembly. Disconnect range gear shift linkage. Support transmission housing and rear axle center housing with suitable safety stands and floor jack or hoist. Remove rear wheels, ROPS roll bar, fenders, seat and seat support, sheet metal panels, step plates and tool box and support. Remove hydraulic suction tube. Disconnect and remove hydraulic pressure tube from diverter valve manifold and hydraulic lift cover. Remove transmission housing top cover and rear axle center housing front cover. Remove cap screws retaining transmission housing to axle center housing and slide transmission forward off locating dowels.

To reinstall transmission, reverse the removal procedure.

OVERHAUL

Model 2120

122. HYDRAULIC SHUTTLE CLUTCH. The hydraulic shuttle clutch (29—Fig. 198) and transmission input shaft (9) can be serviced without removing complete transmission from tractor if desired. Split tractor between engine and clutch housing as outlined in paragraph 87. Remove the steering column

Fig. 198—Hydraulic shift shuttle clutch assembly (29) and transmission drive shaft (9) can be removed after separating shuttle housing from transmission housing.

7. Forward drive gear
8. Spacer
9. Transmission drive shaft
10. Reverse drive gear
29. Forward-reverse clutch pack

with shift levers and the hydraulic shuttle clutch control valve from clutch housing. Disconnect engine clutch control rod. Unbolt and remove hydraulic tubes from front of clutch housing. Remove retaining cap screws, then separate engine clutch housing from shuttle clutch housing.

Unbolt and remove shift arm cover (3—Fig. 199) with shift arm (4), throwout bearing support (13) and adapter housing (38) from clutch housing. Remove step plates from both sides of transmission. Remove top cover from transmission housing. Remove shuttle clutch housing retaining nuts and bolts, then separate clutch housing from transmission housing.

Slide reverse idler gear shaft (42—Fig. 199) out front of shuttle clutch housing while removing idler gear (44), bearing (43) and thrust washer (45).

Withdraw transmission input shaft (9—Fig. 198) with gears (7 and 10) and spacer (8) as an assembly from front of transmission housing. Remove bearings, gears and spacer from input shaft.

Withdraw shuttle clutch assembly (29—Fig. 198) from front of transmission housing. Remove seal rings (35—Fig. 199) from front of clutch shaft. Remove snap ring (34) and withdraw bearing (33), thrust washer (32), reverse gear (31) and thrust washer (22) from front of shaft. Remove snap ring (19), back plate (20) and reverse clutch discs and plates (30) from clutch carrier (29). Compress spring (25), then remove snap ring (23) and washer (24). Remove reverse clutch piston (26) from clutch housing.

Disassemble forward clutch assembly from clutch housing in similar manner as reverse clutch assembly.

NOTE: Remove the small pin (not shown) from clutch shaft prior to removing forward clutch piston.

Inspect all parts for excessive wear, scoring, discoloration from overheating or other damage and renew as necessary. Check thickness of clutch discs and plates. Renew clutch plates and discs as a set if thickness is less than 1.9 mm (0.075 inch). Renew all seal rings and "O" rings.

To reassemble, reverse the disassembly procedure while observing the following instructions. Lubricate all components with clean oil during reassembly. When assembling clutch packs (21 and 30—Fig. 199), alternately install drive plates and clutch friction discs, beginning with a drive plate. Position reverse gear (10) on input shaft (9) with hub facing forward.

Install hydraulic shuttle clutch unit on front of transmission housing. Be sure that two shims are positioned in front of main transmission countershaft front bearing, then install input shaft assembly (9—Fig. 198). Install shuttle clutch housing onto transmission housing and tighten retaining nuts and bolts securely. Install shift arm (4—Fig. 199), making sure lower end of shift arm engages notch in shift rails. Install throwout bearing support (13).

Adjust shuttle clutch shaft bearings as follows: Tap end of clutch shaft (29—Fig. 199) rearward to remove

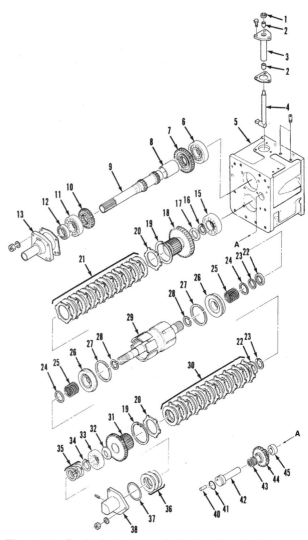

Fig. 199—Exploded view of hydraulic shuttle clutch assembly used on Model 2120.

1. Oil seal	22. Thrust washer
2. Bushings	23. Snap ring
3. Shift cover	24. Washer
4. Shift arm	25. Spring
5. Shuttle housing	26. Clutch piston
6. Bearing	27. Seal ring
7. Forward drive gear	28. Seal ring
8. Spacer	29. Clutch housing
9. Transmission drive shaft	30. Reverse clutch pack
10. Reverse drive gear	31. Reverse counter gear
11. Bearing	32. Thrust washer
12. Oil seal	33. Bearing
13. Clutch release bearing support	34. Snap ring
15. Bearing	35. Seal rings
16. Snap ring	36. Shims
17. Thrust washer	37. "O" ring
18. Forward counter gear	38. Adapter housing
19. Snap ring	40. Pin
20. Back plate	41. "O" ring
21. Forward clutch pack	42. Idler shaft
	43. Needle bearing
	44. Idler gear
	45. Thrust washer

all free play. Measure depth "A" of bearing counterbore in adapter housing (38), and measure distance "B" that clutch shaft front bearing protrudes from front surface of shuttle housing. Subtract measurement "B" from measurement "A." The difference between the two measurements is amount of shims (36) required to provide specified zero preload and zero free play of clutch shaft bearings. Install correct thickness shims and the adapter housing.

123. MAIN TRANSMISSION. Remove transmission assembly from tractor as outlined in paragraph 121. Remove shuttle clutch housing, shuttle clutch assembly and transmission input shaft from front of transmission as outlined in paragraph 122.

Remove shift detent pin (5—Fig. 200), spring (6) and ball (7) from top of transmission housing. Drive roll pin out of first-second (upper) shift fork (2), then slide shift rail (1) out front of housing and remove shift fork. Drive roll pin out of third-fourth (lower) shift fork (4). Turn shift rail (3) 90 degrees, then slide rail out front of housing and remove shift fork, interlock pin (8), ball (7) and spring (6).

Remove snap ring (23—Fig. 201) from rear of transmission housing. Drive transmission countershaft (20) rearward from housing while removing synchronizer assemblies, gears and spacer collar (3 through 19) from front of shaft. Place matching marks on synchronizer sleeve and hub (9) prior to disassembling synchronizer units.

Pull lower countershaft (26) forward and remove shaft with gear set (25) from the housing. Remove

snap ring (38) from rear of housing. Disengage snap ring (32) from its groove and move it forward on rear mainshaft (37). Pull mainshaft rearward from housing while removing thrust washer (28), sliding gear (31), snap ring (32) and sliding gear (33) from shaft.

Remove shift rail lock bolt (15—Fig. 200), rotate range gear shift rail (12) 90 degrees and slide rail out rear of housing. Be careful not to lose detent ball (11) when shift rail is removed. Remove retainer plate (16)

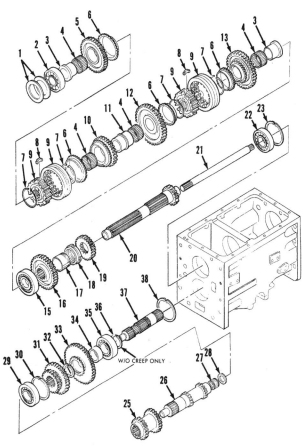

Fig. 201—Exploded view of main and range transmission gears and shafts used on Model 2120 equipped with hydraulic shift shuttle transmission.

1. Shims	21. Pto input shaft
2. Bearing	22. Bearing
3. Collar	23. Snap ring
4. Needle bearing	25. Counter gear,
5. Gear, 3rd	3rd-4th
6. Synchronizer ring	26. Countershaft &
7. Spring	gears, 1st-2nd
8. Key	27. Needle bearing
9. Synchronizer	28. Thrust washer
sleeve & hub	29. Bearing
10. Gear, 4th	30. Snap ring
11. Collar	31. Range sliding
12. Gear, 1st	gear, high-middle
13. Gear, 2nd	32. Snap ring
15. Bearing	33. Range sliding
16. Range gear, high	gear, low
17. Adjustable collar	34. Spacer
18. Nut	35. Bearing
19. Range gear,	36. Snap ring
middle	37. Rear mainshaft
20. Countershaft &	38. Snap ring
low range gear	

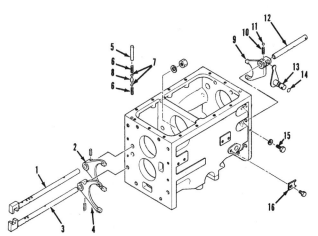

Fig. 200— Exploded view of main and range transmission shift forks and rails used on Model 2120 equipped with hydraulic shuttle transmission.

1. Shift rail	8. Interlock pin
2. Shift fork,	9. Range shift fork
1st-2nd	10. Spring
3. Shift rail	11. Detent ball
4. Shift fork,	12. Shift rail
3rd-4th	13. Shift arm
5. Detent pin	14. "O" ring
6. Springs	15. Retaining screw
7. Detent balls	16. Retaining plate

and withdraw shift arm (13) from housing if necessary.

Clean and inspect components for excessive wear, scoring, chipped teeth or other damage and renew as necessary. Check synchronizer cone wear by measuring clearance (C–Fig. 202) between synchronizer ring (1) and its mating gear (2). Renew synchronizer ring and/or gear if clearance is less than 1.5 mm (0.060 inch). Measure side clearance (D–Fig. 203) between shift forks (1) and their mating sliding collars (2). Renew fork and/or collar if clearance exceeds 1.0 mm (0.039 inch). If any of countershaft components (2 through 20–Fig. 201) were renewed, install a new adjustable collar assembly (17 and 18).

To reassemble, reverse the disassembly procedure while noting the following special instructions. Make certain that needle bearing (27–Fig. 201) is positioned in end of lower countershaft (26) before installing rear mainshaft (37). Start roll pin into hole in lower shift fork (4–Fig. 200) before placing shift fork in housing. Note that roll pin hole in lower shift rail (3) is located further forward than hole in upper shaft rail (1). Install lower shift rail with groove for interlock pin facing upward. Do not drive roll pin into lower shift rail at this time. When assembling synchronizer units, be sure to align sleeve and hub (9–Fig. 201) matching marks made prior to disassembly. Note that shoulder on collar (11) is not centered; longer end of collar must face rearward. The countershaft components cannot be assembled correctly if collar is installed incorrectly. Install main countershaft (20) from the rear while assembling gears and synchronizer units onto front of shaft. Drive roll pin into lower shift fork. Install shift interlock pin (8–Fig. 200), then install upper shift fork, shift rail and detent ball and spring. Install reverse

idler gear (44–Fig. 199) and shaft assembly, making sure that grooved side of thrust washer (45) is facing idler gear.

Install hydraulic shuttle clutch unit on front of transmission housing. Position two shims (1–Fig. 201) in front of countershaft front bearing, then install input shaft assembly (9–Fig. 199). Install shuttle clutch housing onto transmission housing and tighten retaining nuts and bolts securely. Be sure that oil defector trough is positioned inside transmission housing so that it does not come into contact with a gear. Install shift arm (4–Fig. 199), making sure lower end of shift arm engages notch in shift rails. Install throwout bearing support (13). End play of input shaft and countershaft should be 0-0.12 mm (0-0.005 inch). If necessary, install shims (1–Fig. 201) between throwout bearing support (13–Fig. 199) and input shaft front bearing to obtain recommended shaft end play.

Adjust shuttle clutch shaft bearings by means of shims (36–Fig. 199) to provide specified zero preload and zero free play of clutch shaft bearings as outlined in paragraph 122. Install correct thickness shims and the adapter (38).

If transmission case or any parts on input shaft or upper countershaft have been removed, upper shaft gear end play must be adjusted as follows: End play of gears is adjusted by means of an adjustable length collar (17–Fig. 201) and nut (18). A new adjusting collar and nut should be used when adjustment of end play is required. Turn the nut until finger tight while tapping gears forward and rearward. When adjusted correctly, there should be zero preload and zero free play of gears. Make sure that gears and countershaft turn without binding, then stake flange of nut into groove of collar to lock the assembly.

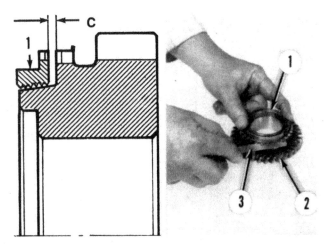

Fig. 202—To check synchronizer cone wear, measure clearance (C) between synchronizer ring (1) and its mating gear (2).

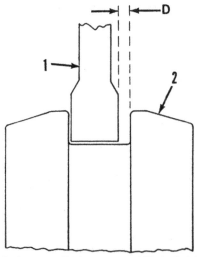

Fig. 203—Clearance (D) between shift fork (1) and shift collar (2) should not exceed 1.0 mm (0.039 inch).

CREEPER GEAR

An optional creeper gear set is available on Models 1920 and 2120 equipped with either 12 × 4 gear transmission or 12 × 12 shuttle transmission. The creeper gear provides an additional 12 forward speeds and four reverse speeds on models equipped with 12 × 4 transmission or an additional 12 forward speeds and 12 reverse speeds on models equipped with 12 × 12 shuttle transmission.

OVERHAUL

Model 1920

124. To disassemble creeper gear, first separate tractor between transmission housing and rear axle center housing as outlined in paragraph 127. Remove transmission top cover. Remove pto coupling (14—Fig. 205) and bearing (12) from front of pto countershaft. Disengage snap ring (16) from its groove, then slide idler gear (15) rearward. Drive pto countershaft forward from axle center housing and lift out idler gear and snap ring. Withdraw sliding gear (2) from bevel pinion shaft.

To remove shift fork (3) and rail (4), remove detent retaining bolt and detent spring and ball (10). Drive roll pins out of shift fork and boss (5), then slide shift rail forward from housing and remove shift fork and boss. Drive pin out of shift lever (9) and remove the lever. Remove retainer plate (8) and withdraw shift arm (6) from housing.

To reassemble, reverse the disassembly procedure.

Model 2120

125. To disassemble creeper gear, first separate tractor between transmission housing and rear axle center housing as outlined in paragraph 127. Remove top cover and hydraulic lift housing from axle center housing. Remove pto rear countershaft as outlined in paragraph 151. If equipped with hydraulic shuttle shift, remove pto one-way clutch assembly (13—Fig. 206). Remove snap ring (14), then drive creeper mainshaft (17) rearward and remove shaft and drive gear (19) out top of housing.

To remove creeper sliding gear (12—Fig. 206), first remove differential lock shaft and fork from axle center housing. Remove differential and ring gear assembly as outlined in paragraph 128. Remove creeper shift rail locking bolt (5). Slide shift rail (6) out front of housing and remove creeper shift fork (7) and front wheel drive shift fork if so equipped. Remove lock nuts from bevel pinion shaft and slide pinion shaft rearward while removing sliding gear from front of shaft.

To reassemble creeper gears, reverse the disassembly procedure. Adjust bevel pinion shaft bearing preload as outlined in paragraph 138.

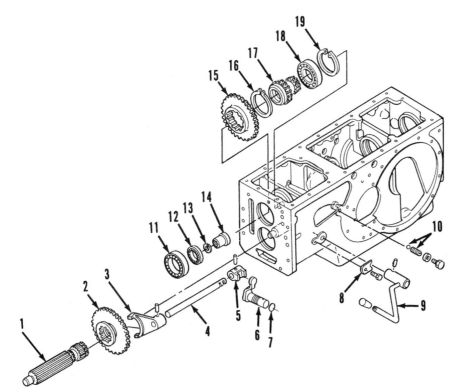

1. Transmission rear mainshaft
2. Creeper sliding gear
3. Shift fork
4. Shift rail
5. Shift boss
6. Shift arm
7. "O" ring
8. Retainer plate
9. Creeper shift lever
10. Detent ball & spring
11. Bearing
12. Bearing
13. Snap ring
14. Coupler
15. Creeper drive gear
16. Snap ring
17. Creeper mainshaft
18. Bearing
19. Snap ring

Fig. 205—Exploded view of optional creeper gear set which is available on Model 1920 equipped with 12 × 4 or 12 × 12 gear transmission.

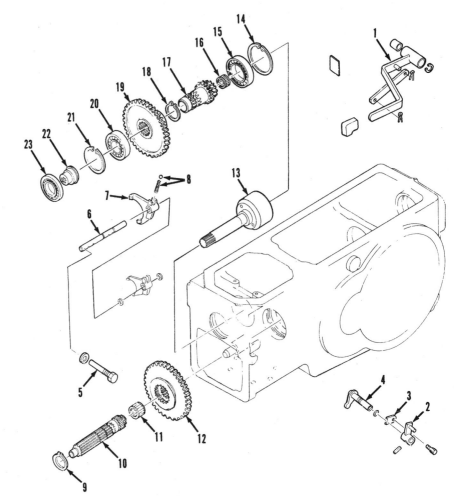

1. Creeper shift linkage
2. Shift lever
3. Retainer plate
4. Shift arm
5. Retainer bolt
6. Shift rail
7. Shift fork
8. Detent ball & spring
9. Snap ring
10. Transmission rear mainshaft
11. Needle bearing
12. Creeper sliding gear
13. Pto one-way clutch
14. Snap ring
15. Bearing
16. Needle bearing
17. Creeper mainshaft
18. Snap ring
19. Creeper drive gear
20. Bearing
21. Snap ring
22. Coupler
23. Bearing

Fig. 206—Exploded view of optional creeper gear set available on Model 2120 equipped with 12 × 4 or 12 × 12 gear transmission. Pto one-way clutch (13) and coupler (22) are used on tractors equipped with hydraulic shuttle shift transmission.

DIFFERENTIAL AND BEVEL DRIVE GEARS

TRACTOR REAR SPLIT

Models 1120-1220

126. To split tractor between rear transmission housing and clutch housing (gear transmission) or extension housing (hydrostatic transmission), first drain oil from transmission housing. If equipped with front wheel drive axle, disconnect front wheel drive shaft. Support the housings separately with suitable safety stands and a rolling floor jack. Remove rear wheels. Remove seat, seat support platform, side panels, step plates, ROPS bar, fenders and transmission cover plate. Disconnect brake rods.

On models with 9 × 3 gear transmission, remove pressure inlet tube (4—Fig. 207) from hydraulic lift housing. Disconnect oil suction tube (2) and oil return tube (3), then unbolt and remove hydraulic filter and manifold (1). Unscrew shift rod turnbuckles

(6) from shift rails, then unbolt and remove shift cover assembly (5).

On models with hydrostatic transmission, remove hydrostatic oil filter outlet tube (4—Fig. 208) and inlet tube (8), hydraulic lift pressure tube (1) and hydraulic system suction tube (9). Unbolt and remove hydrostatic oil filter (5) and hydraulic system oil filter (7) manifolds. Remove hydrostatic return oil tube from left side of extension housing. Remove transmission top cover.

On all models, remove retaining cap screws and roll rear transmission housing away from the tractor.

Models 1320-1520-1720-1920-2120

127. To split tractor between axle center housing and transmission housing, first drain oil from transmission and axle center housings. If equipped with front wheel drive axle, disconnect front wheel drive shaft. Support transmission housing and axle center housing with suitable safety stands. Position wood wedges between front axle and side rails to prevent

Fig. 207—View of hydraulic tubes on Models 1120 and 1220 with gear transmission.

1. Hydraulic oil filter
2. Suction tube
3. Oil return tube
4. Pressure tube
5. Shift cover
6. Turnbuckles

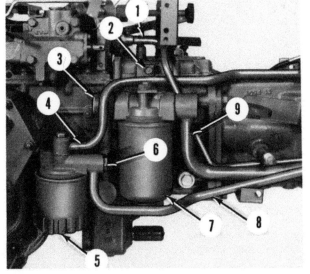

Fig. 208—View of hydraulic tubes on Models 1120 and 1220 with hydrostatic transmission.

1. Pressure tube
2. Filter mounting bracket
3. Suction tube
4. Filter outlet tube
5. Hydrostatic oil filter
6. Banjo bolt
7. Hydraulic system oil filter
8. Filter inlet tube
9. Oil suction tube

tipping. Remove rear wheels. Remove ROPS bar, seat, seat platform, sheet metal panels, step plates, fenders and tool box.

Disconnect both brake rods. Disconnect range gear shift links. Disconnect hydraulic tubes and remove hydraulic system oil filter and mounting flange. Remove high pressure tube from hydraulic lift housing. Remove hydraulic system oil return tube. If equipped with hydrostatic transmission, remove transmission oil return tube from left side of axle center housing.

Remove transmission housing top cover and axle center housing top cover. Attach overhead hoist to axle center housing. Remove retaining cap screws and move axle center housing away from tractor.

DIFFERENTIAL

All Models

128. R&R AND OVERHAUL. The differential assembly may be removed without removing axle center housing from the tractor. Drain oil from transmission and axle housings. Support transmission housing with suitable safety stands. Remove lower lift links. Remove hydraulic lift housing as outlined in paragraph 165. On Models 1320, 1520, 1720, 1920 and 2120, remove differential lock pedal (1—Fig. 210, 211, 212 or 213), shaft (2) and fork (4). On Models 1720, 1920 and 2120, remove pto output shaft bearing retainer from rear of axle center housing. Drive pto countershaft with bearing and rear seal cover rearward from axle center housing. On all models, remove both rear axle housings as outlined in paragraph 140 or 141. Remove differential carrier bearing holders (5 and 27) from both sides of differential, then lift differential assembly out top of housing.

On Models 1120 and 1220, remove cap screws attaching ring gear (10—Fig. 210) to differential case (23). Remove retaining ring (24), then separate pinion shafts (15 and 19), thrust washers (18), pinion gears (16), side gears (13) and thrust washers (12) from case.

On Models 1320 and 1520, remove cap screws attaching bevel ring gear (10—Fig. 211) to differential case (23). On models built prior to November 1989, remove retaining ring (24). On models built in November 1989 and later, drive dowel pin (24A) out of case. On all models, remove pinion shafts (15 and 19), pinion gears (16), thrust washers (18), side gears (13) and thrust washers (12).

On Models 1720, 1920 and 2120, remove carrier bearing (8—Fig. 212 or 213), snap ring (9) and differential lock clutch (6). Remove cap screws retaining end plate (21) to differential case (12) and remove end plate, side gear (14) and thrust washer (13). On Model 2120, unbolt and remove bevel ring gear (11) from differential case. On all models, remove pinion shafts (19 and 20), pinion gears (18) and thrust washers (16) from differential case.

Inspect all parts for excessive wear or damage. Clearance between differential pinion gears and

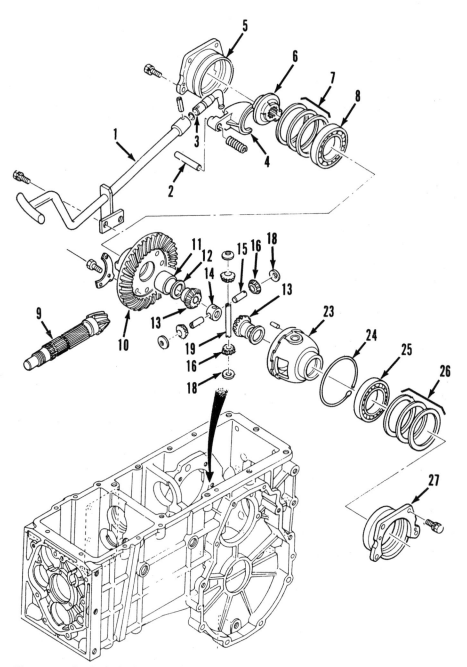

Fig. 210—Exploded view of differential assembly used on Models 1120 and 1220.

1. Differential lock pedal
2. Shaft
3. Shift arm
4. Differential lock shift fork
5. Bearing retainer
6. Differential lock clutch
7. Shims
8. Bearing
9. Bevel pinion shaft
10. Bevel ring gear
11. Bushing
12. Thrust washer
13. Side gear
14. Pinion shaft support
15. Pinion shaft
16. Pinion gear
18. Thrust washer
19. Pinion shaft
23. Differential case
24. Retaining ring
25. Bearing
26. Shims
27. Bearing retainer

shafts should be 0.1 mm (0.004 inch) and maximum allowable clearance is 0.5 mm (0.020 inch). Differential pinion gear thrust washer thickness is 1.2 mm (0.047 inch) when new and wear limit is 0.9 mm (0.035 inch). Specified backlash between differential pinion gears and side gears is 0.10-0.15 mm (0.004-0.006 inch).

To reassemble differential, reverse the disassembly procedure. Install ring gear to differential case using new cap screws and locking plates. On Model 2120, tighten differential case end plate retaining screws to a torque of 30-40 N·m (22-28 ft.-lbs.). Tighten ring gear retaining cap screws to a torque of 26-30 N·m (19-22 ft.-lbs.) on Models 1120 and 1220; 49-

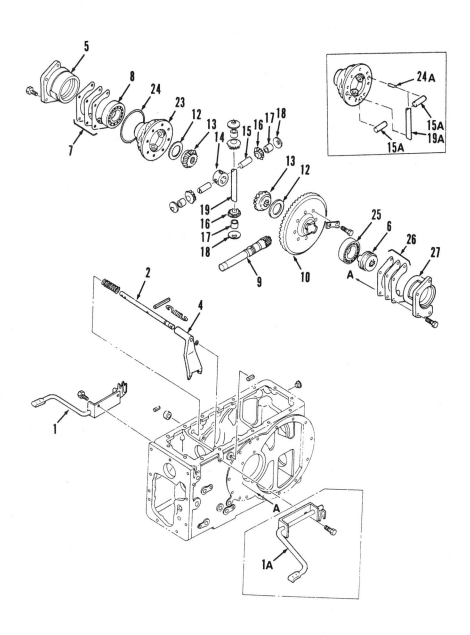

Fig. 211—Exploded view of differential assembly used on Models 1320 and 1520. Pinion shafts (15A and 19A) and retaining pin (24A) are used in tractors built in November 1989 and later. Differential lock pedal (1A) is used on tractors with hydrostatic transmission. Refer to Fig. 210 for legend except for bushings (17).

63 N·m (36-47 ft.-lbs.) on Models 1320, 1520, 1720 and 1920; 60-75 N·m (44-55 ft.-lbs.) on Model 2120.

If original bevel ring gear and pinion, differential case and carrier bearings are being reused, reinstall differential assembly using same shims as removed.

If ring gear, differential case or carrier bearings are being renewed, refer to appropriate paragraph 133 or 139 for adjustment of differential carrier bearings and ring gear and pinion backlash.

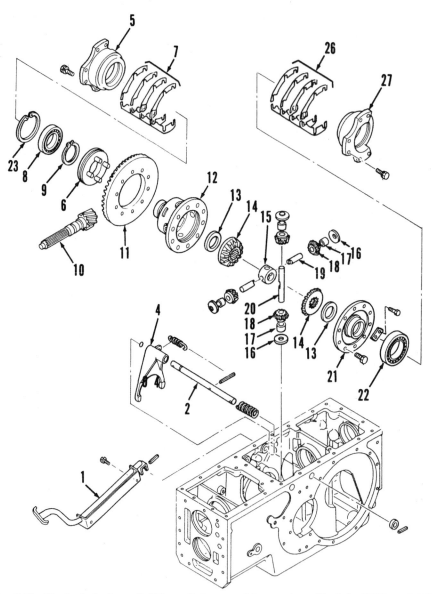

Fig. 212—Exploded view of differential assembly used on Models 1720 and 1920.

1. Differential lock pedal
2. Shift shaft
4. Differential lock shift fork
5. Bearing retainer
6. Differential lock clutch
7. Shims
8. Bearing
9. Snap ring
10. Bevel pinion shaft
11. Bevel ring gear
12. Differential case
13. Thrust washer
14. Side gear
15. Pinion shaft support
16. Thrust washer
17. Bushing
18. Pinion gear
19. Pinion shaft
20. Pinion shaft
21. End cover
22. Bearing
23. Snap ring
26. Shims
27. Bearing retainer

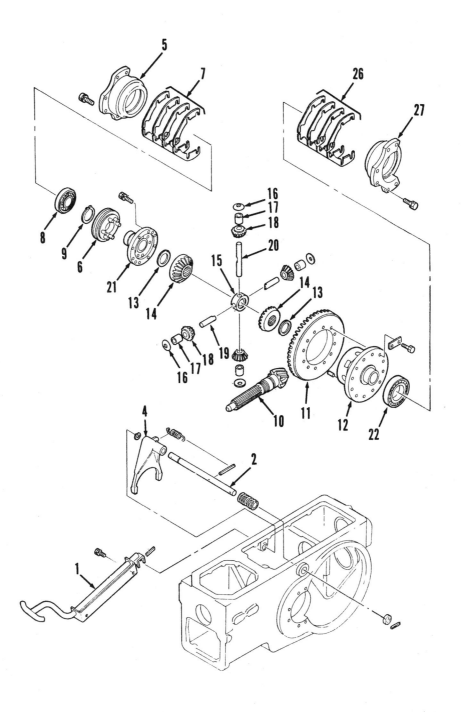

Fig. 213—Exploded view of differential assembly used on Model 2120. Refer to Fig. 212 for legend.

BEVEL DRIVE GEARS

Models 1120-1220

130. REMOVE AND REINSTALL. The bevel pinion and ring gear must be renewed as a matched set. To remove bevel pinion and ring gear, first remove rear transmission housing from the tractor as outlined in paragraph 91. Remove hydraulic lift housing and top cover from transmission housing. Remove differential and ring gear assembly as outlined in paragraph 128. Remove transmission mainshaft and countershaft assemblies as outlined in paragraph 92. Remove locknut (9—Fig. 214) from front of pinion shaft (1), then tap pinion rearward while removing front wheel drive gear (7), if so equipped, snap ring (6) and range sliding gear (5). Remove bearing (4), shims (3) and thrust washer (2) from rear of shaft.

To reinstall bevel pinion and ring gear, reverse the removal procedure while noting the following special instructions. Use new cap screws and lock plates to install ring gear on differential case. Tighten ring gear retaining cap screws to a torque of 26-30 N·m (19-22 ft.-lbs.). Adjust pinion cone point, pinion bearing preload, differential carrier bearings and backlash between ring gear and pinion as outlined in paragraphs 131, 132 and 133.

131. ADJUST BEVEL PINION CONE POINT. If bevel pinion shaft is being reused, original shims (3—Fig. 214) should be installed. If bevel pinion is renewed, required thickness of shims is determined according to the setting numbers stamped on rear face of pinion gears. The setting number will be preceded by a plus (+) or minus (−) sign to indicate the amount of error (in millimeters) from zero adjustment. To determine correct shim thickness, subtract number on new pinion from number on old pinion. The difference between the two numbers is thickness of shims that must be added to or removed from original shim thickness. Example: If old pinion shaft number is +0.1 and new pinion number is -0.1, shim pack thickness must be INCREASED 0.2 mm (0.008 inch). If old pinion number is -0.2 and new pinion number is +0.1, shim pack thickness must be DECREASED 0.3 mm (0.012 inch).

132. ADJUST BEVEL PINION BEARING PRELOAD. Assemble thrust washer, shims and rear bearing on pinion shaft. Insert pinion shaft from the rear while installing front bearing cone, sliding gear, snap ring and front wheel drive gear if so equipped. Install front bearing and a new locknut. Wrap a cord around pinion shaft as shown in Fig. 215 and use a pull scale to measure pull required to rotate shaft. Tighten locknut until a constant pull of 7-9 kg (15-20 pounds) is required to rotate shaft.

133. ADJUST DIFFERENTIAL CARRIER BEARINGS AND BEVEL GEAR BACKLASH. If differential case, carrier bearings, bearing holders and ring gear are all being reused, the original thickness shims (7 and 26—Fig. 210) should be correct to reinstall in original locations. If any of these parts are renewed, carrier bearing adjustment and bevel gear backlash should be adjusted as follows:

Install differential assembly in housing, selecting shims (7 and 26) as required to provide zero side play of carrier bearings. Be sure that there is some backlash between ring gear and pinion when checking differential side play. Do not preload bearings.

After carrier bearings are adjusted, check backlash between bevel gears using a dial indicator located at

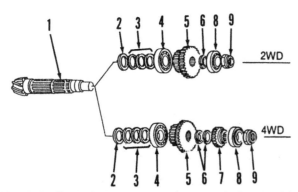

Fig. 214—Exploded view of bevel pinion shaft and related components used on Models 1120 and 1220.

1. Bevel pinion shaft	6. Snap ring
2. Thrust washer	7. Front wheel drive
3. Shims	gear
4. Bearing	8. Bearing
5. Sliding gear	9. Locknut

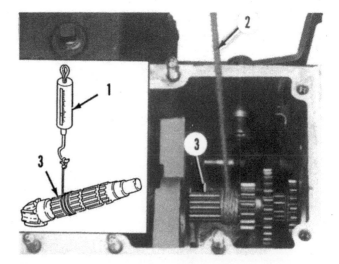

Fig. 215—To adjust pinion shaft bearing preload, wrap a cord (2) around pinion shaft and use a spring scale (1) to measure pull required to rotate shaft. Tighten bearing adjusting nut until specified pull is obtained.

right angle to outer edge of ring gear teeth (Fig. 216). Specified backlash is 0.10-0.15 mm (0.004-0.006 inch). If backlash is excessive, move shims from right side of ring gear to left side. Move shims from left side to right side to increase backlash. Do not add or subtract shims from total shim pack thickness as carrier bearing adjustment would be affected.

Models 1320-1520

135. REMOVE AND REINSTALL. The bevel pinion and ring gear must be renewed as a matched set. To remove bevel gears, first remove rear axle center housing from tractor as outlined in paragraph 127. Remove hydraulic lift housing and top cover from axle center housing. Remove differential and ring gear assembly as outlined in paragraph 128.

Straighten tabs of lockwasher (7—Fig. 217) and loosen pinion shaft nuts (6). Disengage front snap ring (14) from its groove and slide it forward on pinion shaft. Tap pinion shaft rearward from housing while removing front bearing, gears, snap rings, nuts and lockwasher from shaft. Remove rear bearing (4),

shims (3) and thrust washer (2) from shaft if necessary.

To reinstall bevel gears, reverse the removal procedure while noting the following special instructions. Use new cap screws and lock plates to install ring gear on differential case. Tighten ring gear retaining cap screws to a torque of 49-63 N·m (36-47 ft.-lbs.). Adjust pinion bearing preload, differential carrier bearings and backlash between bevel gears as outlined in paragraphs 137, 138 and 139.

Models 1720-1920-2120

136. REMOVE AND REINSTALL. The bevel pinion and ring gear must be renewed as a matched set. To remove bevel gears, first remove rear axle center housing from tractor as outlined in paragraph 127. Remove hydraulic lift housing from axle center housing. Remove differential and ring gear assembly as outlined in paragraph 128.

Straighten tabs of lockwasher (7—Fig. 218) and loosen pinion shaft nuts (6). Tap pinion shaft rearward

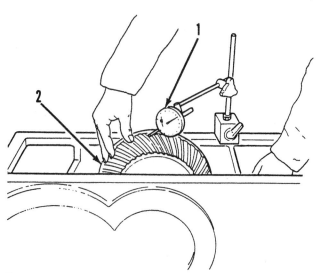

Fig. 216—Use a dial indicator (1) to check backlash between ring gear (2) and bevel pinion gear.

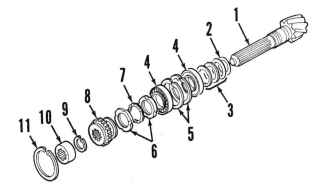

Fig. 218—Exploded view of bevel pinion shaft and related parts used on Models 1720, 1920 and 2120.

1. Bevel pinion shaft
2. Thrust washer
3. Shims
4. Bearings
5. Snap rings
6. Nuts
7. Lockwasher
8. Front wheel drive gear
9. Snap ring
10. Coupling
11. Snap ring

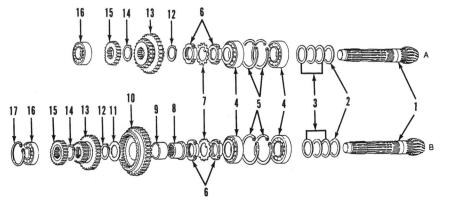

1. Bevel pinion shaft
2. Thrust washer
3. Shims
4. Bearings
5. Snap ring
6. Nuts
7. Lockwasher
8. Coupling
9. Bushing
10. Drive gear
11. Thrust washer
12. Snap ring
13. Range sliding gear
14. Snap ring
15. Front wheel drive gear
16. Bearing
17. Snap ring

Fig. 217—Exploded view of bevel pinion shaft and related parts used on Models 1320 and 1520. View "A" is for tractors equipped with 9 × 3 gear transmission and view "B" is for tractors equipped with hydrostatic transmission.

from housing while removing creeper gear and front wheel drive gear (if so equipped), nuts, lockwasher and front bearing cone from front of shaft. Remove rear bearing (4), shims (3) and thrust washer (2) from shaft if necessary.

To reinstall bevel gears, reverse the removal procedure while noting the following special instructions. Use new cap screws and lock plates to install ring gear on differential case. Tighten ring gear retaining cap screws to a torque of 49-63 N·m (36-47 ft.-lbs.) on Models 1720 and 1920 or 60-75 N·m (44-55 ft.-lbs.) on Model 2120. Adjust pinion bearing preload, differential carrier bearings and backlash between bevel gears as outlined in paragraphs 137, 138 and 139.

Models 1320-1520-1720-1920-2120

137. ADJUST BEVEL PINION CONE POINT. If bevel pinion shaft is being reused, original shims (3—Fig. 217 or 218) should be installed. If bevel pinion is renewed, required thickness of shims is determined according to the setting numbers stamped on rear face of pinion gears. The setting number will be preceded by a plus (+) or minus (−) sign to indicate the amount of error (in millimeters) from zero adjustment. To determine correct shim thickness, subtract number on new pinion from number on old pinion. The difference between the two numbers is thickness of shims that must be added to or removed from original shim thickness. Example: If old pinion shaft number is +0.1 and new pinion number is -0.1, shim pack thickness must be INCREASED 0.2 mm (0.008 inch). If old pinion number is -0.2 and new pinion number is +0.1, shim pack thickness must be DECREASED 0.3 mm (0.012 inch).

138. ADJUST BEVEL PINION BEARING PRELOAD. Assemble thrust washer, shims and rear bearing on pinion shaft. Insert pinion shaft from the rear while installing remainder of components (Fig. 217 or 218) on front of shaft. Wrap a cord around pinion shaft as shown in Fig. 215 and use a pull scale to measure pull required to rotate shaft. Tighten rear nut (6—Fig. 217 or 218) until the following spring scale reading is obtained: 7-9 kg (16-20 pounds) on Models 1320 and 1520, 13-17 kg (29-37 pounds) on Models 1720 and 1920 or 11-13 kg (24-29 pounds) on Model 2120.

After specified bearing preload is obtained, tighten front nut (6) and bend tab of lockwasher (7) into slots in nuts to lock the assembly.

139. ADJUST DIFFERENTIAL CARRIER BEARINGS AND BEVEL GEAR BACKLASH. If differential case, carrier bearings, bearing holders and ring gear are all being reused, the original thickness shims (7 and 26—Fig. 211, 212 or 213) should be

correct to reinstall in original locations. If any of these parts are renewed, carrier bearing adjustment and bevel gear backlash should be adjusted as follows:

Install differential assembly in housing, selecting shims (7 and 26) as required to provide zero side play of carrier bearings. Be sure that there is some backlash between ring gear and pinion when checking differential side play. Do not preload bearings.

After carrier bearings are adjusted, check backlash between bevel gears using a dial indicator located at right angle to outer edge of ring gear teeth (Fig. 216). Specified backlash is 0.10-0.15 mm (0.004-0.006 inch). If backlash is excessive, move shims from left side of ring gear to right side. Move shims from right side to left side to increase backlash. Do not add or subtract shims from total shim pack thickness as carrier bearing adjustment would be affected.

FINAL DRIVE

Models 1120-1220

140. R&R AND OVERHAUL. To remove final drive assembly from either side, first drain oil from transmission and rear axle housings. Remove hitch lower links. Remove rear wheel, ROPS bar and fender. Disconnect brake control rod from brake lever. Attach suitable hoist to rear axle housing, remove retaining cap screws and move axle housing away from tractor.

To disassemble, remove snap ring (7—Fig. 220) and final drive gear (6) from inner end of axle shaft (2). Remove cap screws from seal retainer (1), then drive axle shaft out of housing. Remove seal (3) and bearing (4). Unbolt and remove brake cover (16) and brake shoes (15). Remove snap ring (13) and remove brake drum (12) from final drive pinion shaft (8). Drive pinion shaft and bearing out of axle housing. Remove oil seal (11).

Inspect all parts for excessive wear or damage and renew as necessary. To reassemble, reverse the disassembly procedure.

Models 1320-1520-1720-1920-2120

141. R&R AND OVERHAUL. To remove final drive assembly from either side, first drain oil from rear axle center housing. Remove hitch lower links. Remove rear wheel, ROPS bar and fender. Disconnect brake control rod from brake lever. Support rear axle center housing with suitable stands and attach a hoist to rear axle housing. Remove retaining cap screws and move axle housing away from tractor.

To disassemble, remove locknut (12—Fig. 221), spacer (7) and final drive gear (10) from inner end of axle shaft (1). Remove cap screws from seal retainer (2), then drive axle shaft out of axle housing. Remove oil seal (3) and bearings (5 and 8). Remove brake cover (24) and brake plates (18) and discs (19). Drive final drive pinion (14) out of axle housing.

Inspect all parts for excessive wear or damage and renew as necessary. To reassemble, reverse the disassembly procedure.

BRAKES

ADJUSTMENT

All Models

142. On Models 1120 and 1220, brake pedals should have 20 to 30 mm (3/4 to 1-3/16 inch) free travel (T—Fig. 222). On all other models, recommended pedal free travel is 35 to 45 mm (1-3/8 to 1-3/4 inch).

To adjust pedal free travel, loosen locknut (3) and turn brake control rod (2) as necessary. Shortening the brake rod decreases free travel. Be sure that braking action is equal on both wheels.

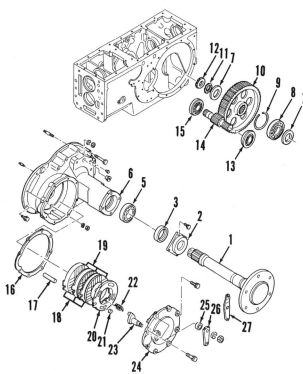

Fig. 221—Exploded view of final drive assembly used on Models 1320, 1520, 1720, 1920 and 2120.

1. Axle shaft	15. Bearing
2. Retainer plate	16. Gasket
3. Oil seal	17. Brake anchor pin
5. Bearing	18. Brake plates
6. Axle housing	19. Brake discs
7. Spacers	20. Brake actuator
8. Bearing	plate
9. Snap ring	21. Actuator ball
10. Final drive gear	22. Spring
11. Lockwasher	23. Brake cam
12. Nut	24. Brake cover
13. Bearing	25. Seal
14. Final drive	26. Brake lever
pinion gear	27. Hand brake lever

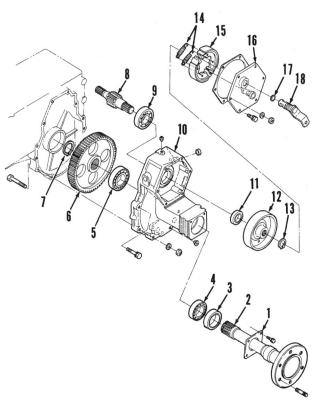

Fig. 220—Exploded view of final drive assembly used on Models 1120 and 1220.

1. Retainer plate	10. Axle housing
2. Axle shaft	11. Oil seal
3. Oil seal	12. Brake drum
4. Bearing	13. Snap ring
5. Bearing	14. Springs
6. Final drive gear	15. Brake shoes
7. Snap ring	16. Brake cover
8. Final drive	17. "O" ring
pinion shaft	18. Brake lever &
9. Bearing	cam

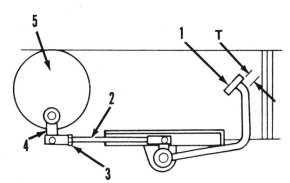

Fig. 222—To adjust brakes, loosen locknut (3) and turn brake rod (2) as necessary until specified pedal free travel (T) is obtained.

Models 1120-1220

143. R&R AND OVERHAUL. To disassemble brakes, first support rear of tractor and remove rear wheels. Disconnect brake rod from brake cover lever. Unbolt and remove brake cover (16—Fig. 220) and brake shoes (15) as an assembly. Remove brake shoes from cover, then remove brake lever and cam (18). Remove snap ring (13) and withdraw brake drum (12) from pinion shaft.

Inspect all parts for excessive wear or other damage and renew as necessary. Inside diameter of brake drum when new is 120 mm (4.725 inches) and maximum allowable inside diameter is 122 mm (4.803 inches).

To reassemble brakes, reverse the disassembly procedure. Adjust brake pedal free play as outlined in paragraph 142.

Models 1320-1520-1720-1920-2120

144. R&R AND OVERHAUL. To disassemble brakes, first support rear of tractor and remove rear wheels. Disconnect brake rod from brake cover lever. Unbolt and remove brake cover (24—Fig. 221), actuator (20), brake plates (18) and discs (19) from axle housing. Remove return springs (22) and separate actuator plate (20) and steel balls (21) from brake cover. Remove brake lever (26) and withdraw brake cam (23). Drive oil seal (25) out of cover.

Inspect all parts for excessive wear or other damage and renew as necessary. Brake discs should be renewed if depth of lining grooves is less than 0.05 mm (0.002 inch). Brake plates should be renewed if they are excessively worn, cracked or warped more than 0.1 mm (0.004 inch).

To reassemble brakes, reverse the disassembly procedure. Adjust brake pedal free travel as outlined in paragraph 142.

POWER TAKE-OFF

STANDARD PTO

Models 1120-1220

A standard 540 rpm pto is available on all tractors. On models with 9 × 3 gear transmission, the pto drive is taken from the transmission input shaft and is transmitted through reduction gears to pto countershaft (3—Fig. 223) and output shaft (28). A one-way clutch (22) is used on the output shaft to transmit power in one direction only and prevent pto equipment from transmitting power back into pto drive line.

On models with hydrostatic transmission, pto drive is taken from hydrostatic input shaft and is transmitted through reduction gears in transmission to pto countershaft (3—Fig. 224) and output shaft (28) located in rear axle center housing. A one-way clutch is not used on tractors with hydrostatic transmission.

145. R&R AND OVERHAUL. The pto output shaft (28—Fig. 225) and related parts can be removed from the rear after removing retaining cap screws from bearing retainer (31). On models with 9 × 3 gear transmission, separate rear bearing retainer (31) and output shaft (28) from front bearing retainer (20) and one-way clutch (22). Remove bearing (29) and oil seal (30) from rear retainer. Remove snap ring (27) and withdraw spacer (26), one-way bearing (25) and needle bearing (23) from one-way clutch housing.

On models with hydrostatic transmission, drive pto output shaft (28—Fig. 225) forward out of bearing re-

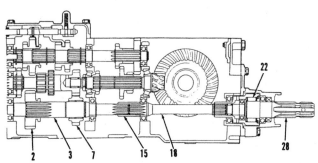

Fig. 223—Cross-sectional drawing of pto drive components used on Models 1120 and 1220 equipped with 9 × 3 gear transmission.

2. Pto sliding gear	15. Coupling
3. Pto front	18. Pto drive shaft
countershaft	22. One-way clutch
7. Idler gear	28. Pto output shaft

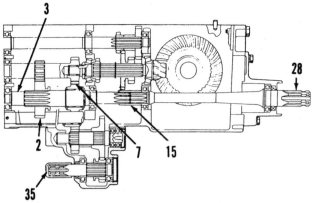

Fig. 224—Cross-sectional drawing of pto drive components used on Models 1120 and 1220 equipped with hydrostatic transmission and mid-mount pto.

2. Pto sliding gear	15. Coupling
3. Pto front	28. Pto output shaft
countershaft	35. Mid-mount pto
7. Idler gear	output shaft

tainer (31). Drive bearing (29) and oil seal (30) out of retainer.

To reassemble output shaft, reverse the disassembly procedure. On gear drive models, be sure that one-way clutch locks up when rear countershaft (18) is rotated clockwise (as viewed from the rear) and rotates freely when countershaft is rotated counterclockwise.

To remove pto front countershaft (3—Fig. 225) and related parts on gear drive models, split tractor between clutch housing and rear transmission as outlined in paragraph 91. Remove transmission mainshaft and countershaft as outlined in paragraph 92. Remove pto output shaft (28) and rear countershaft (18) if not already removed. Disengage snap ring (4) from shaft groove, then drive countershaft rearward until front bearing (1), sliding gear (2), snap ring (4), thrust washers (5) and idler gear (7) with bearings can be removed from front of shaft. Remove countershaft and rear bearing (8) forward from transmission housing.

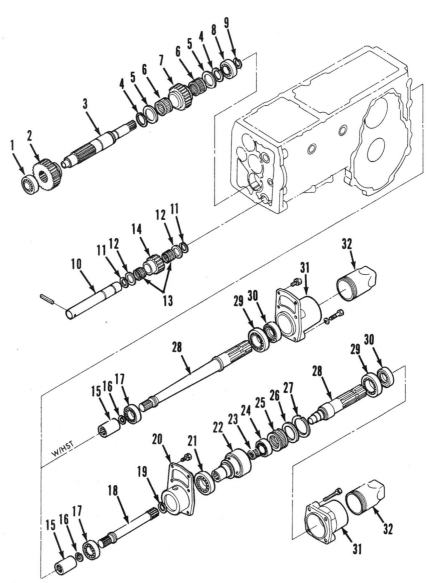

Fig. 225—Exploded view of pto drive components used on Models 1120 and 1220.

1. Bearing
2. Pto sliding gear
3. Front countershaft
4. Snap ring
5. Thrust washer
6. Needle bearings
7. Idler gear
8. Bearing
9. Snap ring

10. Mid-mount pto idler shaft
11. Snap ring
12. Thrust washers
13. Needle bearings
14. Idler gear
15. Coupling
16. Snap ring
17. Bearing

18. Pto drive shaft
19. Snap ring
20. Front bearing retainer
21. Bearing
22. One-way clutch housing
23. Needle bearing
24. Bearing

25. One-way clutch
26. Spacer
27. Snap ring
28. Pto output shaft
29. Bearing
30. Oil seal
31. Rear bearing retainer
32. Guard

To remove pto countershaft (3—Fig. 225) on hydrostatic models, split tractor between extension housing and rear axle center housing and remove range transmission mainshaft and countershaft as outlined in paragraph 105. Remove pto output shaft (28) if not already removed. Disengage snap ring (4) from shaft groove, then drive countershaft rearward until front bearing (1), sliding gear (2), snap ring (4), thrust washers (5) and idler gear (7) with bearings (6) can be removed from front of shaft. Remove countershaft with rear bearing out front of housing.

MID-MOUNT PTO

Models 1120-1220

146. R&R AND OVERHAUL. An optional mid-mount pto is available on all tractors. To remove mid-

mount pto gearbox, first disconnect shift control linkage from shifter arm. Remove gearbox retaining cap screws and remove gearbox from bottom of transmission housing.

To disassemble gearbox, pry plug (32—Fig. 226 or 227) out of case and remove snap ring (30). Remove snap ring (18) from groove in countershaft (28). Thread an 8 mm bolt into end of countershaft and pull countershaft rearward from case. Pry plug (16) from rear of case and remove snap ring (15). Drive output shaft (12) and rear bearing (14) out rear of case. Remove front bearing (11) and oil seal (10) from case. Remove detent ball and spring (4). Drive roll pin out of shifter arm (8) and remove shift lever (5) and arm. Drive oil seal (7) out of the case.

To remove mid-mount pto countershaft (10—Fig. 225) and gear (14) on gear drive tractors, split tractor between clutch housing and rear transmission as

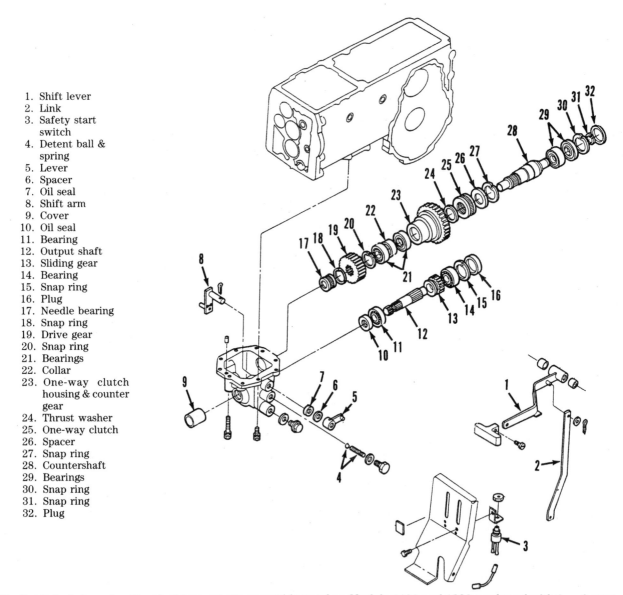

1. Shift lever
2. Link
3. Safety start switch
4. Detent ball & spring
5. Lever
6. Spacer
7. Oil seal
8. Shift arm
9. Cover
10. Oil seal
11. Bearing
12. Output shaft
13. Sliding gear
14. Bearing
15. Snap ring
16. Plug
17. Needle bearing
18. Snap ring
19. Drive gear
20. Snap ring
21. Bearings
22. Collar
23. One-way clutch housing & counter gear
24. Thrust washer
25. One-way clutch
26. Spacer
27. Snap ring
28. Countershaft
29. Bearings
30. Snap ring
31. Snap ring
32. Plug

Fig. 226—Exploded view of optional mid-mount pto assembly used on Models 1120 and 1220 equipped with 9 × 3 gear transmission.

outlined in paragraph 91. Remove transmission main-shaft and countershaft as outlined in paragraph 92. Remove pto front countershaft (3) as outlined in paragraph 145. Remove snap rings (11), then slide shaft (10) forward from housing and remove idler gear (14) assembly.

To remove mid-mount pto countershaft (10—Fig. 225) and gear (14) on hydrostatic models, split tractor between extension housing and rear axle center housing as outlined in paragraph 105. Remove range transmission mainshaft and countershaft as outlined in paragraph 105. Remove pto front countershaft (3) as outlined in paragraph 145. Remove snap rings (11), then slide shaft (10) forward and remove idler gear (14) assembly.

To reassemble, reverse the disassembly procedure.

STANDARD PTO

Models 1320-1520

A single-speed (540 rpm), transmission drive pto is standard on Models 1320 and 1520. The pto is driven from transmission input shaft through reduction

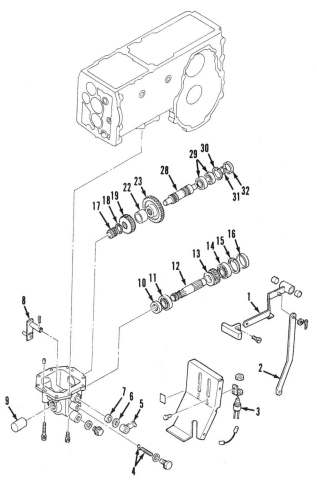

Fig. 227—Exploded view of optional mid-mount pto assembly used on Models 1120 and 1220 equipped with hydrostatic transmission. Refer to Fig. 226 for legend.

gears in transmission housing on models with 9×3 gear transmission. On tractors equipped with hydrostatic transmission, pto is driven by input shaft of hydrostatic unit through reduction gears in transmission housing. A live pto is optional on models with 9×3 gear transmission. Pto is driven by a second clutch disc on engine clutch through a separate pto input shaft.

A one-way clutch is used on models with single disc clutch and 9×3 gear transmission to transmit power in one direction only and prevent pto equipment from transmitting power back into pto drive line. One-way clutch is not used on models with hydrostatic transmission or models with dual disc clutch, live pto.

147. R&R AND OVERHAUL. Pto output shaft (20—Fig. 230) can be removed after first removing hydraulic lift housing as outlined in paragraph 165. Remove retaining cap screws from retainer plate (25), then drive output shaft out the rear and remove spacer (18) and sliding gear (19) through top of housing.

Tractor must be split between transmission and rear axle center housing as outlined in paragraph 127 to remove remainder of pto drive components. On models with 9×3 transmission and single plate clutch, slide one-way clutch (3—Fig. 230) off the rear countershaft (11). On models with 9×3 transmission and dual plate clutch (live pto), slide rear drive shaft (3A) off rear countershaft (11). On all models, remove snap ring (2), then drive rear countershaft (11), bearing (12) and seal cover (13) rearward from housing. Remove drive gear (9) and front bearing (6). To remove pto front countershaft, refer to appropriate transmission disassembly procedure, paragraph 95 for 9×3 gear transmission or paragraph 106 for hydrostatic transmission.

To reinstall pto shafts and gears, reverse the removal procedure. If equipped with one-way clutch, make sure that one-way clutch locks when pto rear countershaft is turned counterclockwise (viewed from rear) and turns freely when shaft is turned clockwise.

MID-MOUNT PTO

Models 1320-1520

148. R&R AND OVERHAUL. An optional mid-mount pto (Fig. 231) is available on all models. The mid-mount pto is driven by a countershaft (4) that is in constant mesh with rear pto countershaft (5).

To remove mid-mount pto gearbox (2), first drain oil from transmission housing. Disconnect shift control linkage from shifter arm. Remove gearbox retaining cap screws and remove gearbox from bottom of transmission housing.

To remove mid-mount pto countershaft (4), first remove pto output shaft (7), one-way clutch or drive shaft (8) and rear pto countershaft (5) as outlined in paragraph 147. Remove seal cover (6) and snap ring from rear of axle center housing. On tractors built prior to February 1988, remove snap ring from front of countershaft. On tractors built in February 1988 and after, remove nut from front of countershaft. On all tractors, drive countershaft rearward from housing and remove drive gear (9).

To disassemble pto gearbox, remove retaining bolt and pull shift rail (9—Fig. 232) out rear of case. Re-

move shift fork (7). Remove shaft safety cap (11) and sealing plug (18) from case. Remove snap ring (17), then drive pto output shaft (15) out rear of case. Remove bearing (13) and oil seal (12).

To reassemble, reverse the disassembly procedure.

STANDARD PTO

Models 1720-1920

A transmission driven 540 rpm pto is standard equipment on Model 1720 equipped with 12 × 4 gear transmission. A one-way clutch is used on output shaft of transmission driven pto to prevent pto equipment from transmitting power back into pto drive line. A live 540 rpm pto is optional equipment on Model 1720 equipped with 12 × 4 gear transmission. A live 540 rpm pto is standard equipment on Model 1720 equipped with 12 × 12 shuttle transmission and on Model 1920 equipped with 12 × 12 shuttle transmission and 12 × 4 gear transmission. A two-speed (540 and 1000 rpm) pto is available as optional equipment on tractors equipped with live pto.

150. R&R AND OVERHAUL. To remove pto components, split tractor between transmission and rear axle center housing as outlined in paragraph 127 and remove hydraulic lift housing as outlined in paragraph 165. Remove output shaft seal retainer (28—

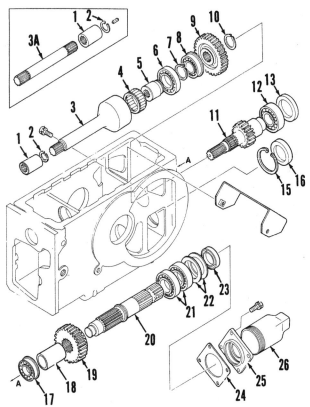

Fig. 230—Exploded view of pto drive components used on Models 1320 and 1520. One-way clutch (3 and 4) is used on tractors with 9 × 3 gear transmission and single disc engine clutch. Drive shaft (3A) is used in place of one-way clutch on tractors equipped with hydrostatic transmission or 9 × 3 transmission with dual disc engine clutch (live pto).

1. Coupling	12. Bearing
2. Snap ring	13. Seal cover
3. One-way clutch	15. Snap ring
housing	16. Seal cover
3A. Drive shaft	17. Bearing
4. One-way clutch	18. Spacer
5. Inner race	19. Drive gear
6. Bearing	20. Pto output shaft
7. Snap ring	21. Bearings
8. Bearing	22. Shims
9. Pto sliding gear	23. Oil seal
10. Snap ring	24. Gasket
11. Pto rear	25. Retainer plate
countershaft	26. Cover

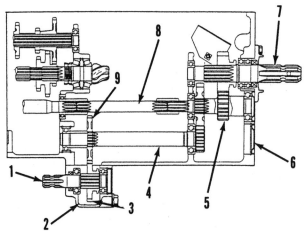

Fig. 231—Cross-sectional drawing of mid-mount pto and rear pto drive components used on Models 1320 and 1520.

1. Mid-pto output	
shaft	
2. Mid-pto gearbox	
3. Sliding gear	6. Seal cover
4. Mid-pto	7. Rear pto output
countershaft	shaft
5. Rear pto	8. Pto drive shaft
countershaft	9. Mid-pto drive gear

Fig. 234) from rear of axle center housing. Drive countershaft (5), rear bearing (6) and seal cover (7) rearward from housing. Remove coupling (2) and bearing (8) from front of housing. Remove snap ring (10) from front of output shaft (25). Drive output shaft rearward from housing and remove one-way clutch assembly (12 through 20) on tractors with transmission drive pto or sliding gear (18A or 18B) on tractors with live pto. To remove pto input shaft, refer to appropriate paragraph 109 or 114 covering overhaul of transmission.

To reinstall pto components, reverse the removal procedure.

STANDARD PTO

Model 2120

A live pto is standard equipment on Model 2120 tractors equipped with 12 × 4 gear transmission. The pto drive is taken from front disc of dual disc engine clutch. On tractors equipped with 12 × 12 shuttle transmission, the pto drive is taken from transmission input shaft which is splined to pto front countershaft. A one-way clutch is used on tractors with 12 × 12 shuttle transmission to prevent pto equipment from transmitting power back into pto drive line.

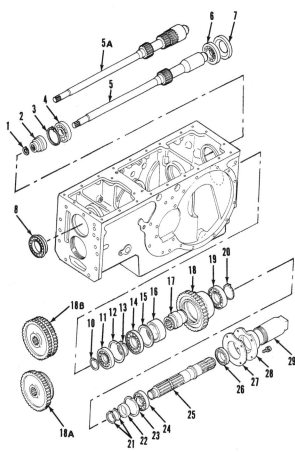

Fig. 234—Exploded view of transmission driven pto components used on Model 1720 with nonsynchromesh 12 × 4 transmission. Models 1720 and 1920 equipped with live pto are similar except that sliding gear (18A or 18B) is used in place of one-way clutch assembly (12 through 20). Countershaft (5A) and sliding gear (18B) are used on tractors equipped with two-speed pto.

1. Snap ring	16. One-way clutch
2. Coupling	17. Coupling
3. Snap ring	18. Clutch hub &
4. Bearing	sliding gear
5. Pto countershaft	19. Bearing
5A. Countershaft	20. Snap ring
(two-speed pto)	21. Snap rings
6. Bearing	22. Spacer
7. Seal cover	23. Snap ring
8. Bearing	24. Bearing
10. Snap ring	25. Pto output shaft
11. Bearing	26. Oil seal
12. Snap ring	27. Gasket
13. Snap ring	28. Seal retainer
14. Bearing	29. Cover
15. Thrust washer	

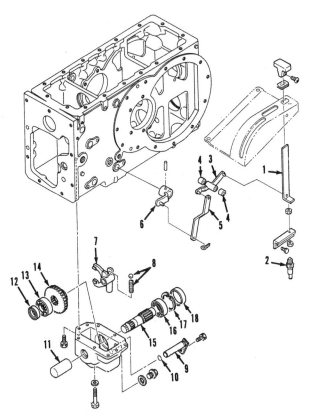

Fig. 232—Exploded view of mid-mount pto assembly used on Models 1320 and 1520.

1. Shift lever	9. Shift rail
2. Safety start	10. "O" ring
switch	11. Cover
3. Shift lever	12. Oil seal
4. Bushings	13. Bearing
5. Link	14. Sliding gear
6. Shift lever	15. Output shaft
7. Shift fork	16. Bearing
8. Detent ball &	17. Snap ring
spring	18. Seal cover

151. R&R AND OVERHAUL. To remove pto components, split tractor between transmission and rear axle center housing as outlined in paragraph 127. Remove hydraulic lift housing as outlined in paragraph 165 and top cover from axle center housing. Remove countershaft seal cover (17—Fig. 235) and snap ring (16) from rear of housing. Drive pto countershaft (14) and bearing (15) rearward from housing. Remove output shaft seal retainer (29), pull output shaft out rear of housing and remove sliding gear (20). Remove one-way clutch assembly (1 through 10), if so equipped, from front of axle center housing.

To reinstall pto components, reverse the removal procedure. If equipped with one-way clutch, be sure that clutch locks up when clutch hub (6—Fig. 235) is rotated counterclockwise (viewed from rear) and rotates freely when turned clockwise.

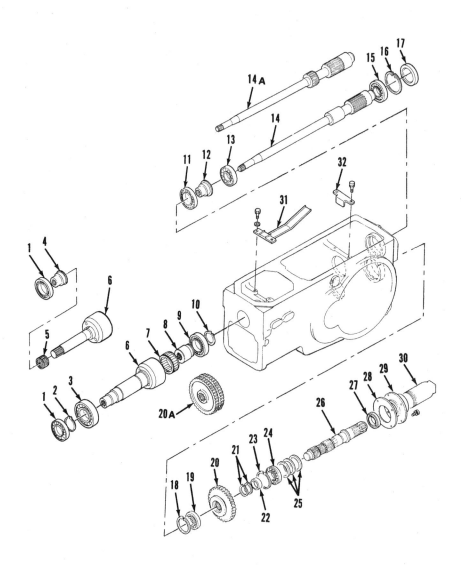

Fig. 235—Exploded view of pto drive components used on Model 2120. One-way clutch assembly (1 through 10) is used on tractors equipped with shuttle transmission. Coupling (4) and clutch hub (6) are used on tractors with shuttle transmission and creep gear. Countershaft (14A) and sliding gear (20A) are used on tractors equipped with two-speed pto.

1. Bearing		18. Snap ring	25. Shims
2. Snap ring	12. Coupling	19. Bearing	26. Pto output shaft
3. Bearing	13. Bearing	20. Sliding gear	27. Oil seal
4. Coupling	14. Pto counter-	(single-speed)	28. Gasket
5. Needle bearing	shaft (single-	20A. Sliding gear	29. Seal retainer
6. Clutch hub	speed)	(two-speed)	30. Cover
7. One-way clutch	14A. Countershaft	21. Snap rings	31. Oil trough (w/
8. Coupling	(two-speed)	22. Spacer	shuttle
9. Bearing	15. Bearing	23. Snap ring	transmission)
10. Snap ring	16. Snap ring	24. Bearing	32. Bearing retainer
11. Bearing	17. Seal cover		

HYDRAULIC LIFT SYSTEM

The hydraulic lift system consists basically of an oil reservoir, hydraulic pump, control valve, lift cylinder and lift links. A single lever, position control hydraulic lift system is standard equipment on all models except Model 2120. A two lever, position and draft control system is optional on Models 1320, 1520, 1720 and 1920 and standard on Model 2120.

FLUID AND FILTER

All Models

152. The transmission housing and rear axle center housing serve as a common oil reservoir for the hydraulic system. It is recommended that the oil be drained every 300 hours of operation and refilled with new Ford 134 oil or equivalent fluid. Capacity is approximately 16 liters (16.9 U.S. quarts) on Models 1120 and 1220, 22 liters (23.3 U.S. quarts) on Models 1320 and 1520, 27 liters (28.5 U.S. quarts) on Model 1720, 29 liters (30.6 U.S. quarts) on Model 1920 and 33 liters (34.9 U.S. quarts) on Model 2120.

The hydraulic system filter should also be renewed after every 300 hours of operation. The filter is located on right side of transmission housing.

TROUBLE-SHOOTING

All Models

153. The following are symptoms which may occur during operation of the hydraulic lift system and their possible causes. Use this information in conjunction with TESTING and ADJUSTMENT information when diagnosing hydraulic lift problems.

1. Hitch will not lift load. Could be caused by:
   a. Restricted hydraulic oil filter.
   b. Linkage out of adjustment or broken.
   c. System relief valve pressure setting too low.
   d. Safety relief valve faulty.
   e. "O" ring failure between control valve and valve cover.
   f. Oil leakage past lift cylinder piston seal.
   g. Unload valve faulty (1120 and 1220).
   h. Oil leakage past lowering valve poppet.
2. Hitch will not lower. Could be caused by:
   a. Flow control valve in closed position.
   b. Lowering valve spool out of adjustment.
3. Hitch will not raise to full height. Could be caused by:
   a. Position control valve spool out of adjustment.

4. System relief valve opens when hitch is in full raise position. Could be caused by:
   a. Position control valve spool out of adjustment.
5. Lift arms cycle up and down "hiccup" when control lever is in neutral. Could be caused by:
   a. Lift check valve faulty.
   b. Oil leakage past main control valve spool.
   c. Oil leakage past lowering valve.
   d. Oil leakage past lift cylinder seal ring

TESTING

All Models

154. RELIEF VALVE PRESSURE. To check system relief valve opening pressure, remove plug (8—Fig. 236 or 237) from system relief valve-diverter manifold (2) and install a 0-35000 kPa (0-5000 psi) pressure gage. Start the engine and operate hydraulic system until oil is at normal operating temperature. Operate engine at high idle speed. Use a screwdriver to rotate diverter valve spool (5) counterclockwise and observe pressure gage reading.

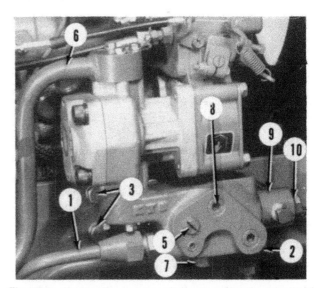

Fig. 236—Hydraulic system relief valve-diverter manifold is mounted below the hydraulic pump on Models 1120 and 1220.

1. Pressure line	
2. Relief valve-	
diverter manifold	
3. Mounting bolts	7. Relief valve plug
5. Diverter valve	8. Pressure port
spool	plug
6. Pump suction	9. Oil return line
line	10. Banjo bolt

Relief Valve Opening Pressure
Models 1120-1220 . . . . .12275-13240 kPa
                                    (1780-1920 psi)
Models 1320-1520 . . . . .10590-11570 kPa
                                    (1535-1680 psi)
Model 1720 . . . . .11965-12940 kPa
                                    (1735-1875 psi)
Model 1920 . . . . .14205-15170 kPa
                                    (2060-2200 psi)
Model 2120 . . . . .16685-17650 kPa
                                    (2420-2560 psi)

If necessary, add or remove shims (9—Fig. 238 or
239) to adjust pressure setting. A change in shim
thickness of 0.1 mm (0.004 inch) will change pressure
setting approximately 260 kPa (38 psi) on Models 1120
and 1220 or approximately 350 kPa (50 psi) on all other models.

SYSTEM RELIEF AND DIVERTER VALVE ASSEMBLY

All Models

155. R&R AND OVERHAUL. The combination
system relief-diverter valve manifold (2—Fig. 236 or
237) is located in the high pressure line on right side
of tractor. The valve assembly contains hydraulic system relief valve (7) which protects hydraulic pump
from excessive pressure. The valve also contains connection ports for use of auxiliary hydraulic equipment and a diverter valve spool (5) which allows
selection of operation of either auxiliary hydraulics
or hitch lift system.

**Fig. 237—Hydraulic system relief valve-diverter manifold
is mounted on right side of tractor on Models 1320, 1520,
1720, 1920 and 2120.**

1. Pressure line	
2. Relief valve- diverter manifold	6. Pump suction line
3. Mounting bolts	7. Relief valve plug
4. Pump pressure line	8. Pressure port plug
5. Diverter valve spool	9. Oil return line 10. Banjo bolt

To remove valve assembly, disconnect hydraulic
tubes from valve body. Remove valve retaining cap
screws and remove valve assembly.

Remove relief valve plug (10—Fig. 238 or 239),
shims (9), spring (7) and relief valve and seat (6).
Drive roll pin (2) out of diverter valve spool (3) and
withdraw spool from valve body.

Inspect all parts for excessive wear, scoring or other damage and renew as necessary. Renew valve as
an assembly if diverter valve spool or body is damaged. Reassemble valve using new "O" rings (1 and
8).

HYDRAULIC PUMP

A gear type hydraulic pump is used on all models.
Rated pump capacity at 2500 engine rpm is 17.6
liters/minute (4.6 gallons/minute) on Models 1120 and
1220, 24.9 liters/minute (6.5 gallons/minute) on
Models 1320 and 1520, 29.6 liters/minute (7.8 gal-
lons/minute) on Models 1720 and 1920, and 35
liters/minute (9.2 gallons/minute) on Model 2120.

All Models Except 2120

156. REMOVE AND REINSTALL. The hydraulic
system pump (2—Fig. 240) is mounted on right side
of the engine and driven by engine timing gears. To

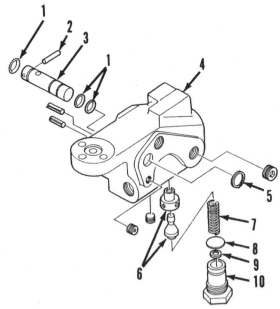

**Fig. 238—Exploded view of relief valve-diverter manifold
assembly used on Models 1120 and 1220.**

1. "O" rings	6. Relief valve
2. Pin	poppet & seat
3. Diverter valve spool	7. Spring
4. Manifold	8. "O" ring
5. Snap ring	9. Shim 10. Plug

remove the hydraulic pump, first remove side screen and lower panel. Clean the pump, fittings and surrounding area to prevent entry of dirt into hydraulic system. Disconnect suction tube manifold (1) from pump housing. On Models 1120 and 1220, disconnect

high pressure tube (3) from diverter valve manifold (4) and remove cap screws attaching pump to manifold. On all other models, disconnect high pressure tube from pump housing. On all models, remove mounting bolts and withdraw pump from engine timing gear case.

To reinstall pump, reverse the removal procedure.

Model 2120

157. REMOVE AND REINSTALL. To remove the hydraulic pump (2—Fig. 241), drain coolant from radiator and engine. Remove hood assembly, side screen, lower panel and front grille. Remove the air cleaner tube and the battery and battery support. Disconnect radiator hoses. Remove radiator support brace and radiator mounting nuts, then remove radiator assembly. Clean pump, fittings and surrounding area to prevent entry of dirt into hydraulic system. Disconnect suction tube (3) and pressure tube (4) from hydraulic pump. Remove pump mounting cap screws and withdraw pump assembly from front of engine timing gear case.

To reinstall pump, reverse the removal procedure.

All Models

158. OVERHAUL. The only parts serviced in hydraulic pump are the shaft seal and internal seal rings and back-up rings. If pump is excessively worn or damaged, renew pump as an assembly.

To disassemble pump, first scribe match marks on end cover and body to ensure correct alignment

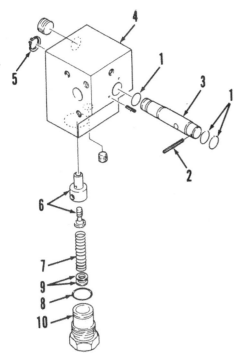

Fig. 239—Exploded view of relief valve-diverter manifold assembly used on Models 1320, 1520, 1720, 1920 and 2120. Refer to Fig. 238 for legend.

Fig. 240—View of hydraulic system pump (2) used on Models 1120 and 1220. Models 1320, 1520, 1720 and 1920 are similar except that relief valve-diverter manifold (4) is not mounted beneath the pump.

1. Suction line	4. Relief valve-diverter manifold
2. Hydraulic system pump	
3. Pressure line	5. Power steering pump

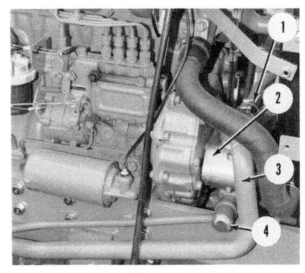

Fig. 241—Hydraulic system pump is mounted on front of engine timing gear housing on Model 2120.

1. Suction line (w/hydraulic shift shuttle)
2. Hydraulic system pump
3. Suction line
4. Pressure line

HITCH ADJUSTMENTS

when reassembling. Remove cover retaining screws, then carefully separate end cover from pump body. Identify all parts as they are removed so that they can be reinstalled in their original positions. Refer to appropriate Fig. 242 or 243 for exploded view of pump components.

Lubricate all parts during assembly with clean hydraulic oil. On Models 1120, 1220, 1320 and 1520, be sure that rectangular slots in gear wear plates (6—Fig. 242) are positioned on pressure side of pump and that wear plate with two holes is installed on end cover side of gears. On Models 1720, 1920 and 2120, be sure that dowel pins (7—Fig. 243) are in place between the gear bearings (10). Install seals (6) with flat side outward into bearing grooves. Install back-up rings (5) into the step on inside of the seals.

All Models

160. POSITION CONTROL ROD. The length of position control rod (2—Fig. 245 or 246) must be adjusted correctly to provide proper operation of hitch. If rod is too short, the control valve spool will not return to neutral when lift arms reach maximum lift position and system relief valve will be actuated. If control rod is too long, full lift height will not be obtained.

On Models 1120 and 1220, loosen locknuts (3—Fig. 245) and remove control rod (2) from lift arm. On all

Fig. 245—View of position control linkage used on Models 1120 and 1220.

2. Position control rod
3. Locknuts
4. Control link
5. Position control lever

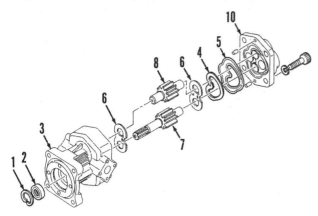

Fig. 242—Exploded view of hydraulic system pump used on Models 1120, 1220, 1320 and 1520.

1. Snap ring
2. Seal
3. Pump housing
4. Back-up ring
5. Seal ring
6. Wear plates
7. Drive gear
8. Idler gear
10. End cover

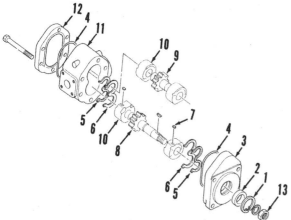

Fig. 243—Exploded view of hydraulic system pump used on Models 1720, 1920 and 2120.

1. Snap ring
2. Seal
3. Pump flange
4. Seal ring
5. Back-up ring
6. Seal
7. Key
8. Drive gear
9. Idler gear
10. Bearings
11. Pump housing
12. End cover
13. Nut

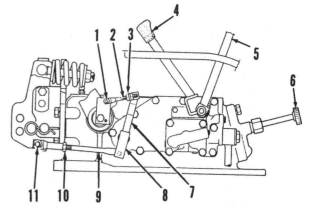

Fig. 246—Drawing of position control and draft control linkage used on Models 1320, 1520, 1720, 1920 and 2120.

1. Lift arm pin
2. Position control rod
3. Locknut
4. Draft control lever
5. Position control lever
6. Flow control valve
7. Position control arm
8. Draft control arm
9. Draft control rod
10. Locknut
11. Top link pin

other models, loosen locknut (3—Fig. 246) and remove pin (1) from lift arm link. On all models, start the engine and move position control lever (5) to full raise position. Lift arms should raise and relief valve should open. Move control lever downward until control valve spool is returned to neutral position and relief valve is closed. Adjust length of position control rod (2) to align control rod with hole in lift arm, then lengthen control rod one additional turn. Connect control rod to lift arm and tighten locknut. Check operation of hitch to make certain that relief valve does not operate when lift arms reach the top of their travel.

161. LEVER FRICTION. The control lever (1—Fig. 247) should require 2-3 kg (4-7 pounds) of pull at lever knob to move the lever. Lever friction is adjusted by removing cotter pin (2) and turning nut (3) until desired lever friction is obtained.

Models With Draft Control

162. DRAFT CONTROL ROD. After position control rod is adjusted correctly, adjust draft control rod (9—Fig. 246) as follows: Move position control lever (5) to highest lifting position, but not in notch at upper end of quadrant. Move draft control lever to upper end of quadrant. Loosen locknut (10) on draft control rod, then disconnect control rod from draft control arm pin (11). Turn flow control knob (6) to full open position. Start engine and run at 1000-1500 rpm. Move draft control rod rearward until system goes on high pressure and relief valve opens, then slowly move rod forward until relief valve closes. Adjust length of control rod to align with draft control arm pin, then shorten rod one more turn. Connect rod to draft control rod and tighten locknut.

To check the adjustment, start the engine and move draft control and position control levers to full raise position. Then move position control lever down to lower position. If lift arms do not lower or if they lower too slowly, shorten draft control rod one more turn.

163. TOP LINK MAIN SPRING. Length (L—Fig. 248) of top link main spring (3) should be 96 mm (3-3/4 inches) on Models 1320 and 1520 or 77 mm (3 inches) on Models 1720, 1920 and 2120. To adjust spring length, remove cotter pin and turn adjusting nut (2) as necessary.

164. CONTROL LEVER NEUTRAL ADJUSTMENT. Start engine and operate at 1000-1500 rpm. Move position control lever (1—Fig. 249) rearward to approximately the center of quadrant and scribe a line on quadrant at rear edge of control lever. Slowly move lever forward until lift arms just start to lower. Lever travel distance (N) should be 10 mm (7/16 inch).

To adjust control lever neutral travel, the control valve assembly must be removed from lift cover as outlined in paragraph 170. If travel is excessive, turn control valve adjusting screw (2) clockwise to decrease travel distance. If lever travel is too small, turn adjusting screw counterclockwise. Note that insufficient lever neutral travel will cause hitch to "hunt" for neutral position.

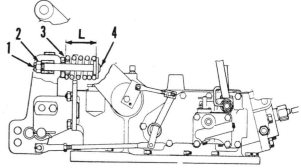

Fig. 248—On tractors equipped with draft control, length (L) of top link main spring (3) should be adjusted to 96 mm (3-3/4 inches) on Models 1320 and 1520 or 77 mm (3 inches) on Models 1720, 1920 and 2120.

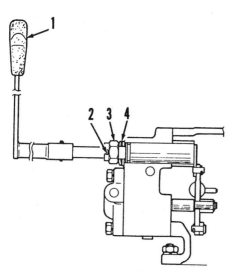

Fig. 247—To adjust control lever friction, remove cotter pin (2) and turn adjusting nut (3) to compress spring washers (4).

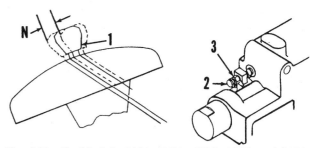

Fig. 249—On Models 1320, 1520, 1720, 1920 and 2120, position control lever neutral travel (N) should be 10 mm (7/16 inch). Refer to text for adjustment.

HYDRAULIC LIFT HOUSING

All Models

165. REMOVE AND REINSTALL. Completely lower lift arms to remove oil from lift cylinder. Remove rear wheels, roll bar, fenders, seat, seat support and sheet metal panels. Disconnect draft control feedback rod, if so equipped, and remove top link bracket as an assembly from lift housing. Disconnect lift links from lift housing arms. Remove shift levers and support bracket from lift housing. Disconnect high pressure line from lift housing. Remove lift cover retaining bolts and nuts, noting that bolts of different lengths are used at various locations. Use a suitable hoist to remove lift housing from tractor.

To reinstall lift housing, reverse the removal procedure. Adjust hitch control linkage as outlined in paragraphs 160 through 164.

166. OVERHAUL LIFT CYLINDER AND ROCKSHAFT. Prior to removing lift arms (11—Fig. 250), scribe reference marks across lift arms and rockshaft (15) and across ram arm (8) and rockshaft to ensure correct alignment when reassembling. Disconnect position control feedback rod from lift arm. Remove snap rings (10) and pull lift arms off rockshaft. Slide rockshaft out left side of lift housing while sliding ram arm off the shaft.

Unbolt and remove cylinder head (1), then push piston (4) and cylinder liner (3) out front of lift housing.

Inspect all parts for excessive wear or damage and renew as necessary. When installing new "U" shaped seal (5) on piston, be sure lip of seal (open side) is positioned toward front of piston. When renewing rockshaft bushings, drive bushings into housing to the following specified depth (D—Fig. 251). On Models 1120 and 1220, left bushing (1) should be 6.0 mm (0.236 inch) below flush and right bushing (2) should be 6.5 mm (0.256 inch) below flush. On Models 1320 and 1520, both bushings should be 6 mm (0.236 inch) below flush. On Models 1720, 1920 and 2120, both bushings should be 7 mm (0.276 inch) below flush. Note that inside diameter of right bushing is smaller than inside diameter of left bushing. Renew all "O" rings and seals.

Lubricate parts with clean hydraulic oil during assembly. Be sure to align reference marks made prior to disassembly on ram arm, lift arms and rockshaft. Tighten cylinder head retaining cap screws to a torque of 75-88 N·m (55-65 ft.-lbs.).

FLOW CONTROL VALVE, CHECK VALVE AND SAFETY RELIEF VALVE

All Models

167. During the "raise" cycle, pressure oil from hydraulic pump flows past the flow control valve needle and seat (1—Fig. 252), opens the check valve (2) and enters lift cylinder to raise the lift arms. During the "lower" cycle, the check valve is closed, forcing all return oil from cylinder to flow past the flow control valve needle and seat. Thus, lift arm lowering speed is controlled by adjusting flow control valve needle to seat clearance. When flow control valve is

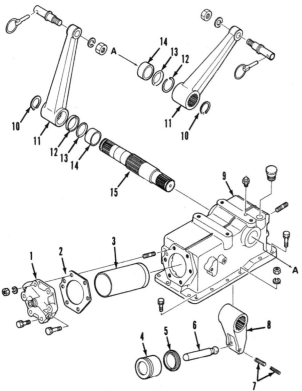

Fig. 250—Exploded view of hydraulic lift housing components.

1. Cylinder head	
2. Gasket	9. Lift housing
3. Cylinder liner	10. Snap ring
4. Piston	11. Lift arms
5. Seal	12. Spacer
6. Piston rod	13. "O" ring
7. Pins	14. Bushings
8. Ram arm	15. Rockshaft

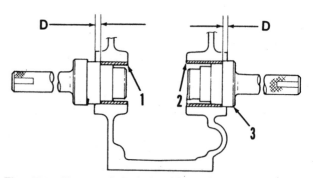

Fig. 251—New rockshaft bushings (1 and 2) must be installed to correct depth (D) in lift housing. Refer to text.

fully closed, oil is trapped in lift cylinder and lift arms will remain at their set height.

The safety relief valve (3) protects the lift cylinder against shock loads when the main control valve is in "neutral" position. When lift cylinder pressure exceeds 22540-26460 kPa (3270-3835 psi) on Models 1120 and 1220, 23000-27000 kPa (3335-3915 psi) on Models 1320, 1520, 1720 and 1920 or 22500-26500 kPa (3270-3840 psi) on Model 2120, the safety relief valve opens and trapped oil in lift cylinder flows directly to sump.

To disassemble, unscrew guide nut (5—Fig. 253) from cylinder head (8), then screw flow control valve stem (6) out of cylinder head. Unscrew safety relief valve body (17) from cylinder head and remove check valve assembly (11, 12, 13 and 14). Remove spring guide bolt (22), shims (21), spring (19) and relief valve ball (18) from valve body, making certain that shims are retained for reassembly.

Inspect all parts for excessive wear or damage and renew as necessary. Reassemble using new "O" rings and seals.

LIFT CONTROL VALVE

Models 1120-1220

168. The control valve assembly contains the unload valve (4—Fig. 254), control valve spool (5), lowering valve (6 and 7) and load check valve (8).

In "neutral" position, pump oil flow is blocked by the control valve spool (5). Oil pressure on front side of unload valve (4) opens the valve and pump oil flow returns to sump (1). The lowering valve (6) and load check valve (8) are both closed by spring pressure, trapping oil in the lift cylinder (11).

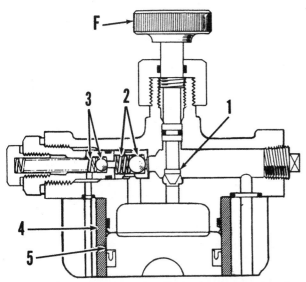

Fig. 252—Cross section of cylinder head showing flow control valve, check valve and safety relief valve.

F. Flow control knob
1. Flow control valve
2. Check ball & spring
3. Safety relief valve assy.
4. Lift cylinder liner
5. Piston

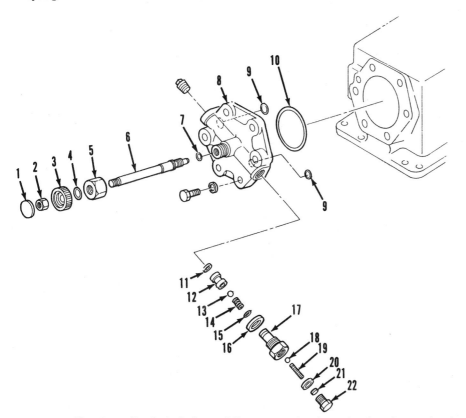

1. Cover plate
2. Nut
3. Knob
4. "O" ring
5. guide nut
6. Flow control valve needle
7. "O" ring
8. Cylinder head
9. "O" ring
10. "O" ring
11. Gasket
12. Check valve seat
13. Check ball
14. Spring
15. "O" ring
16. Seal ring
17. Spring guide & relief valve seat
18. Safety valve ball
19. Spring
20. Seal washer
21. Shim
22. Plug

Fig. 253—Exploded view of flow control valve, check valve and safety relief valve components.

In "raise" position, control valve spool (5) is pulled outward. Pump oil flow is directed to both sides of unload valve (4) and to front side of load check valve (8). Spring pressure holds the unload valve on its seat, blocking oil return passage to sump. When pump pressure overcomes lift cylinder pressure on back side of load check valve, the check valve opens and oil flows to lift cylinder to raise the lift arms. The lift arms will raise until position control feedback linkage returns the control valve spool to neutral position.

In "lower" position, control valve spool (5) is moved inward, blocking oil passage to back side of unload valve (4) and moving lowering valve spool (6) off its seat (7). Pump pressure opens unload valve and pump flow returns to sump. Oil in lift cylinder flows past the unseated drop poppet valve and returns to sump. The lift arms will lower until feedback linkage returns the control valve spool and lower valve spool to neutral position.

169. R&R AND OVERHAUL. The control valve assembly is mounted on outside of lift housing on right side. To remove valve, first fully lower lift arms to release trapped oil. Disconnect position control feedback rod from control valve lever. Remove position control lever retaining nuts and remove lever assembly. Unbolt and remove control valve assembly from lift housing.

To disassemble valve, drive roll pin out of control valve spool and remove control lever and link from valve spool. Unbolt and remove end cover (1—Fig. 255) and plate (20).

NOTE: Before removing lowering valve spool (24—Fig. 256), measure distance (D) from face of locknuts to end of valve spool to facilitate reassembly.

Remove locknuts (5—Fig. 255) and push lowering valve spool (24), valve seat (25) and plug (27) out of valve body. Remove retaining nut (3), spring washers (4) and plate (6) from control valve spool (13), then withdraw spool from valve body. If necessary, remove spool clevis (17), bypass spool (15) and spring (14) from control valve spool. Remove unload valve plug (8), spring (10) and poppet (11). Remove check valve plug (33), spring (31), check valve (30) and seat (29).

Inspect valve spools and valve body bores for scratches, scoring or excessive wear and renew as necessary. Inspect unload valve, check valve and lowering valve seats for wear or damage. Be sure to renew all "O" rings.

To reassemble control valve, reverse the disassembly procedure while noting the following special instructions. If original lowering valve components are reused, adjust lowering valve spool locknuts to distance (D—Fig. 256) as determined before disassembly. If distance (D) is not known or if lowering valve was renewed, adjust the lowering spool as follows:

Install locknuts loosely so that distance (D) is less than 14 mm (0.550 inch). Reinstall valve assembly on lift housing. Start engine and move position control lever to fully raise the lift arms. Stop the engine and slowly move control lever forward until lift arms start to lower. This is "neutral" position. Remove end cover and adjust locknuts (5—Fig. 256) so that clearance (C) between plate (6) and first locknut is 0.3 mm (0.012 inch). Tighten locknuts to secure the adjustment.

Models 1320-1520-1720-1920-2120

170. The control valve assembly contains control valve spool (3—Fig. 257), bypass plunger (4), lowering valve (8 and 9) and load check valve (10).

1. Reservoir
2. Hydraulic pump
3. Main relief valve
4. Unload valve
5. Control valve spool
6. Lowering valve poppet
7. Lowering valve seat
8. Check valve
9. Flow control valve
10. Safety relief valve
11. Lift cylinder

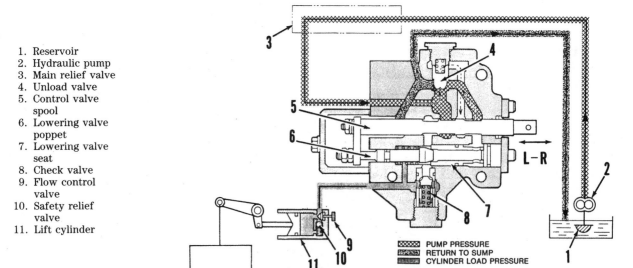

PUMP PRESSURE
RETURN TO SUMP
CYLINDER LOAD PRESSURE

Fig. 254—Cross-sectional drawing of lift control valve used on Models 1120 and 1220. Control valve is shown in neutral position.

In "neutral" position, control valve spool (3) is centered in valve body. Pressure oil from pump flows through an orifice in valve spool to rear face of bypass plunger (4). Oil pressure moves the plunger and aligns oil passages (5) in valve spool and plunger. Pump oil flow passes through the passages and returns to sump (R). The lowering valve spool (9) and check valve (10) are spring loaded closed, trapping oil in lift cylinder (7).

In "raise" position, control valve spool (3) moves inward closing oil passages in valve spool and bypass plunger. Oil then flows around control valve spool and lowering valve spool (8) to front side of load check valve (10). When pump pressure exceeds lift cylinder pressure on back side of check valve, the check valve is lifted off its seat and oil flows to lift cylinder to raise lift arms. The lift arms will raise until feedback linkage returns control valve spool to neutral position.

In "lower" position, control valve spool (3) is moved outward, aligning passages in valve spool and bypass plunger and allowing pump flow to return to sump (R). At the same time, the adjusting bolt attached to valve spool pin contacts lowering valve (9) and pushes it off its seat. Oil in lift cylinder flows past the lowering valve spool and seat and returns to sump, allowing lift arms to lower. The lift arms lower until feedback linkage returns control valve spool to neutral position.

171. R&R AND OVERHAUL. The control valve assembly (15—Fig. 257 or 258) is mounted on the valve cover (4) on right side of lift housing. To remove valve, first fully lower lift arms to discharge trapped oil from valve. Remove right rear wheel, ROPS roll bar and right fender. Disconnect high pressure oil line from valve cover. Disconnect position control follow-up rod and draft control follow-up rod (if so equipped). Unbolt and remove valve cover with control valve from lift housing. Remove cap screws retaining control valve to valve cover and remove valve assembly.

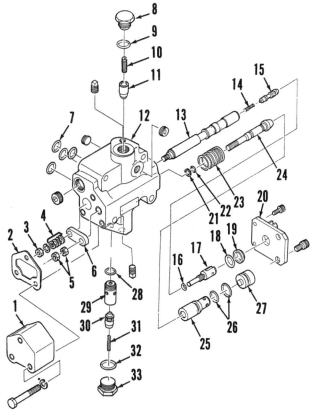

Fig. 255—Exploded view of lift control valve assembly used on Models 1120 and 1220.

1. End cover	18. "O" ring
2. Gasket	19. Wiper seal
3. Nut	20. End plate
4. Spring washers	21. Back-up ring
5. Locknuts	22. "O" ring
6. Valve plate	23. Spring
7. "O" ring	24. Lowering valve
8. Plug	spool
9. "O" ring	25. Lowering valve
10. Spring	seat
11. Unload valve	26. "O" rings
12. Control valve	27. Plug
body	28. "O" ring
13. Control valve	29. Check valve seat
spool	30. Check valve
14. Spring	poppet
15. Bypass valve	31. Spring
spool	32. "O" ring
16. "O" ring	33. Plug
17. Clevis	

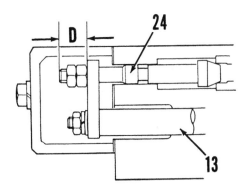

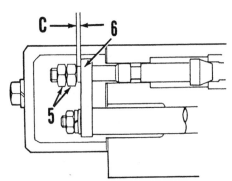

Fig. 256—Measure dimension (D) before disassembling lowering valve spool (24) from valve plate. Refer to text for adjustment of clearance (C).

Refer to Fig. 260 for an exploded view of control valve assembly. To disassemble, remove end cap (1) and spring (2) from valve body. Remove screw (20), plug (22) and adjusting screw pin (26) from valve spool, then pull spool from valve body.

NOTE: Do not disturb setting of adjusting screw (24) unless necessary. If removal of adjusting screw is required, measure distance that adjusting screw extends from the pin (26) prior to removal so that screw can be reinstalled to original setting.

Remove plug (23) from end of spool. Remove cotter pin and washer from pin (32), push inward on spring seat (31) and remove the pin. Withdraw spring seat, spring (30) and bypass plunger (29) from valve spool. Remove plug (12), collar (13), spring seat (14), spring (15) and lowering valve assembly (16 and 18). Remove plug (11), spring seat (9), springs and check valve assembly (4 and 6).

Inspect all parts for excessive wear, scratches, scoring or other damage and renew as necessary. Control valve spool and body must be renewed as a complete assembly. Renew all "O" rings and seals.

Lubricate all valve components with clean hydraulic oil during reassembly. Tighten check valve plug (11) to a torque of 58 N·m (43 ft.-lbs.). Tighten lowering valve plug (12) to a torque of 24 N·m (18 ft.-lbs.). Reinstall control valve and valve cover on lift housing. Adjust position control linkage and draft control linkage (if so equipped) as outlined in paragraphs 160 through 164.

REMOTE CONTROL VALVE

All Models So Equipped

Single spool and double spool remote control valves are available as optional accessories. The single spool

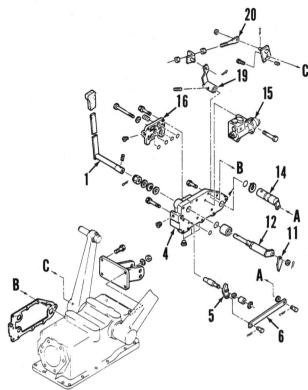

Fig. 258—Exploded view of lift control linkage used on Models 1320, 1520, 1720, 1920 and 2120 equipped with single lever position control.

1. Position control lever
4. Control valve cover
5. Feedback cam
6. Link
11. Position control cam
12. Position control shaft
14. Feedback shaft
15. Control valve assy.
16. Valve cover
19. Feedback arm
20. Position control feedback rod

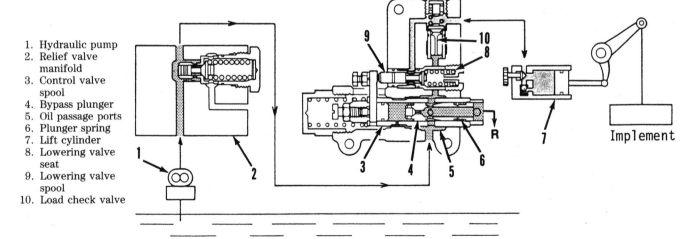

1. Hydraulic pump
2. Relief valve manifold
3. Control valve spool
4. Bypass plunger
5. Oil passage ports
6. Plunger spring
7. Lift cylinder
8. Lowering valve seat
9. Lowering valve spool
10. Load check valve

Fig. 257—Cross-sectional drawing of lift control valve assembly used on Models 1320, 1520, 1720, 1920 and 2120. Control valve is shown in neutral position.

control valve is mounted on right side of hydraulic lift cover. The double spool control valve is mounted on right side of the hood.

172. SINGLE SPOOL VALVE. To remove valve, disconnect hydraulic tubes from valve. Remove control lever from valve spool. Remove retaining cap screws and withdraw valve assembly from lift housing.

To disassemble, unbolt and remove end cap (1—Fig. 261) and control lever bracket (11). Pull valve spool (6) assembly from valve body. Remove retaining bolt (2) and separate centering spring (4) and seats (3) from spool. Remove oil seals (7) from valve body bore.

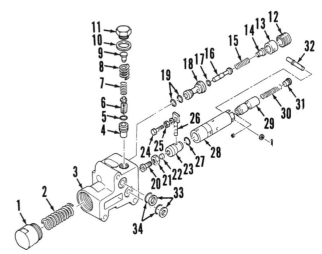

Fig. 260—Exploded view of lift control valve assembly used on Models 1320, 1520, 1720, 1920 and 2120.

1. End cap	18. Lowering valve
2. Centering spring	seat
3. Control valve	19. "O" rings
body	20. Bolt
4. Check valve seat	21. Locknut
5. "O" ring	22. Plug
6. Check valve	23. Plug
poppet	24. Adjusting screw
7. Spring	25. Locknut
8. Spring	26. Adjusting screw
9. Spring seat	pin
10. Gasket	27. "O" ring
11. Plug	28. Control valve spool
12. Plug	29. By-pass plunger
13. Collar	30. Spring
14. Spring seat	31. Spring seat
15. Spring	32. Pin
16. Lowering valve	33. Back-up rings
spool	34. "O" rings
17. "O" ring	

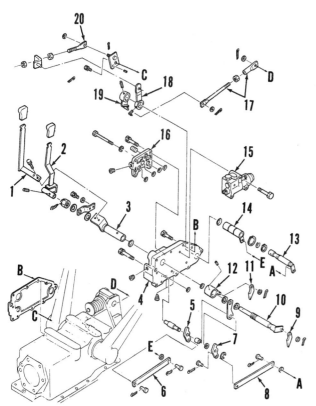

Fig. 259—Exploded view of lift control linkage used on Models 1320, 1520, 1720, 1920 and 2120 equipped with position control and draft control.

1. Position control lever	
2. Draft control lever	12. Position control arm
3. Position control shaft	13. Draft control feedback shaft
4. Control valve cover	14. Position control feedback shaft
5. Position control feedback cam	15. Control valve assy.
6. Feedback link	16. Cover plate
7. Draft control feedback cam	17. Draft control feedback rod
8. Feedback cam	18. Position control arm
9. Draft control cam	19. Draft control arm
10. Draft control shaft	20. Position control feedback rod
11. Position control cam	

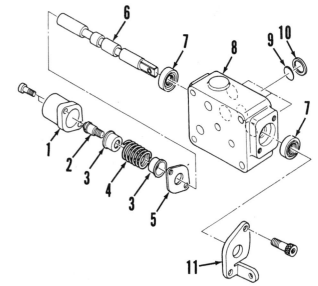

Fig. 261—Exploded view of single spool hydraulic remote control valve.

1. End cap	
2. Retaining bolt	7. Oil seals
3. Spring seats	8. Valve body
4. Centering spring	9. "O" ring
5. Seal retainer plate	10. Back-up ring
6. Valve spool	11. Control lever bracket

Inspect valve spool and bore for pitting, scratches or excessive wear. Make certain that valve spool moves freely in valve body bore. Valve body and spool are available only as an assembly.

Install new oil seals in body bore. Lubricate spool with clean hydraulic oil and reassemble in valve body. Reinstall valve assembly using new "O" ring (9) and back-up ring (10).

173. DOUBLE SPOOL VALVE. To remove valve, remove sheet metal cover and disconnect hydraulic tubes from valve. Disconnect control levers from valve spools. Unbolt and remove valve assembly from mounting bracket.

To disassemble, unbolt and remove seal retainer plates (1—Fig. 262) and end caps (12 and 22). Be careful not to lose detent balls (17) when end cap (22) is removed. Withdraw spools (3 and 4) from valve body. Remove retaining bolts (11 and 15) and separate centering springs and seats from valve spools. Remove oil seals (2) from valve body bores. Remove check valves (24 through 27). On Models 1320, 1520 and 1720, unscrew plug (35) and remove relief valve assembly (28 through 36).

Inspect spools and valve bores for scratches, excessive wear or other damage. Make certain that valve spools move freely in their bores. Spools and valve body must be renewed as an assembly. Check orifice (23) to make sure that it is not plugged. Renew oil seals (2) and "O" rings.

Lubricate all parts with clean hydraulic oil during reassembly. Reinstall valve assembly on tractor. On Models 1320, 1520 and 1720, check and adjust remote valve pressure relief valve setting as outlined in paragraph 174.

174. REMOTE VALVE PRESSURE TEST. The double spool remote valve used on Models 1320, 1520 and 1720 is equipped with a pressure relief valve. On all other models, remote control valve relief pressure is controlled by hydraulic lift system main relief valve.

To check remote control valve relief pressure, connect a 0-35000 kPa (0-5000 psi) pressure gage to remote valve "raise" pressure outlet. Start engine and operate at 2500 rpm. Move remote valve control lever to "raise" position and note pressure gage reading.

Pressure should be 12275-13240 kPa (1780-1920 psi) on Models 1120 and 1220; 14205-15170 kPa (2060-2200 psi) on Model 1920 and 16685-17650 kPa (2420-2560) on Model 2120. If necessary, adjust pressure setting by adding or removing shims in main relief valve.

Pressure should be 10590-11570 kPa (1535-1675 psi) on Models 1320 and 1520 or 11965-12940 kPa (1735-1875 psi) on Model 1720. To adjust remote valve relief pressure, loosen locknut (36—Fig. 262) and turn adjusting screw (33) as required.

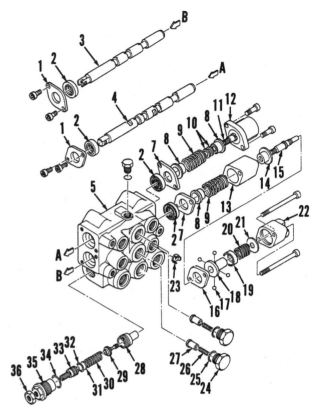

Fig. 262—Exploded view of double spool hydraulic remote control valve. Note that valve spools (3 and 4) are not interchangeable. Pressure relief valve (28 through 36) is used on Models 1320, 1520 and 1720.

1. Seal retainer plate
2. Oil seals
3. Valve spool
4. Valve spool
5. Valve body
7. Seal retainer plate
8. Spring seats
9. Centering spring
10. Shims
11. Retaining bolt
12. End cover
13. Spacer
14. Spring seat
15. Retaining bolt
16. Retaining plate
17. Detent balls
18. Detent guide
19. Collar
20. Spring
21. Washer
22. End cover
23. Orifice
24. Plug
25. "O" ring
26. Spring
27. Check valve poppet
28. Relief valve seat
29. Relief valve
30. Spring
31. "O" ring
32. Back-up ring
33. Adjusting screw
34. "O" ring
35. Plug
36. Locknut

175. Refer to appropriate Fig. 263 through 266 for tractor wiring diagrams. Refer to paragraphs 76 through 85 for service procedures covering electrical system components.

WIRING DIAGRAMS

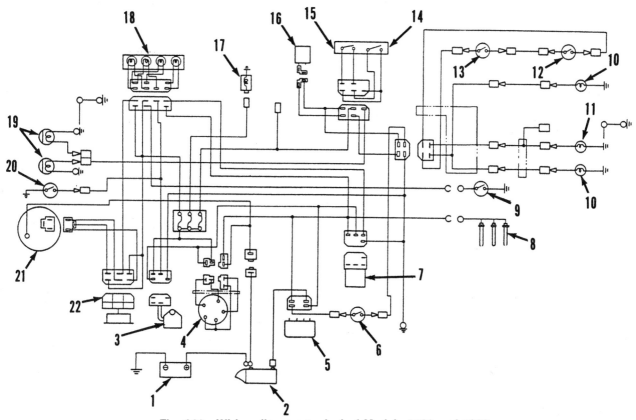

Fig. 263—Wiring diagram typical of Models 1120 and 1220.

1. Battery
2. Starter motor
3. Fuse box
4. Key switch
5. Relay
6. Safety start switch (clutch)
7. Glow plug timer
8. Glow plugs
9. Coolant switch
10. Hazard warning light
11. Tail light
12. Safety switch (mid pto)
13. Safety switch (rear pto)
14. Light switch
15. Hazard switch
16. Flasher unit
17. Fuel shut-off solenoid
18. Instrument panel lights
19. Headlights
20. Oil pressure switch
21. Alternator
22. Regulator

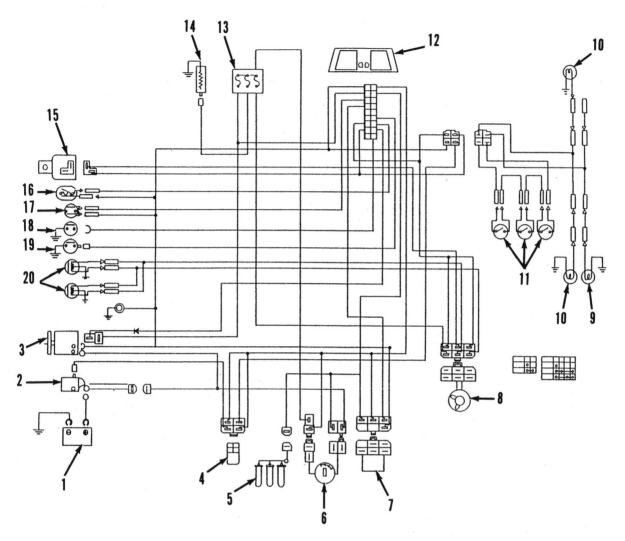

Fig. 264—Wiring diagram typical of Models 1320 and 1520.

1. Battery
2. Starter motor
3. Alternator
4. Neutral start switch
5. Glow plugs
6. Key switch
7. Glow plug timer relay
8. Light switch
9. Tail light
10. Hazard warning lights
11. Neutral start switches
12. Instrument panel lights
13. Fuse box
14. Fuel shut-off solenoid
15. Flasher unit
16. Fuel sender unit
17. Air cleaner sender unit
18. Oil pressure sender unit
19. Temperature sender unit
20. Headlights

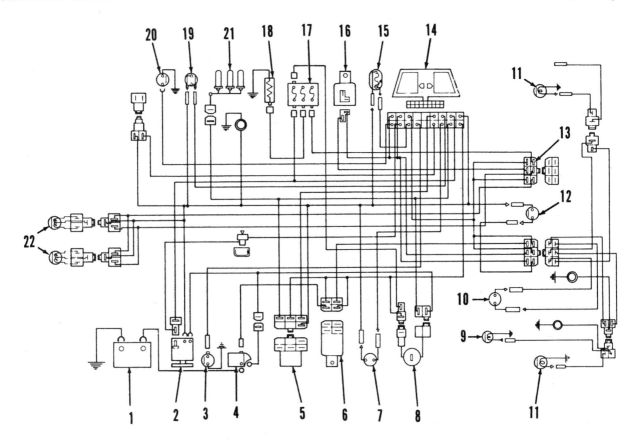

Fig. 265—Wiring diagram typical of Models 1720 and 1920.

1. Battery
2. Alternator
3. Temperature sender unit
4. Starter motor
5. Glow plug timer
6. Neutral start switch
7. Parking brake switch
8. Key switch
9. Tail light
10. Neutral start switch
11. Hazard warning lights
12. Neutral start switch
13. Light switch
14. Instrument panel lights
15. Fuel sender unit
16. Flasher unit
17. Fuse box
18. Fuel shut-off solenoid
19. Air cleaner sender unit
20. Oil pressure sender unit
21. Glow plugs
22. Headlights

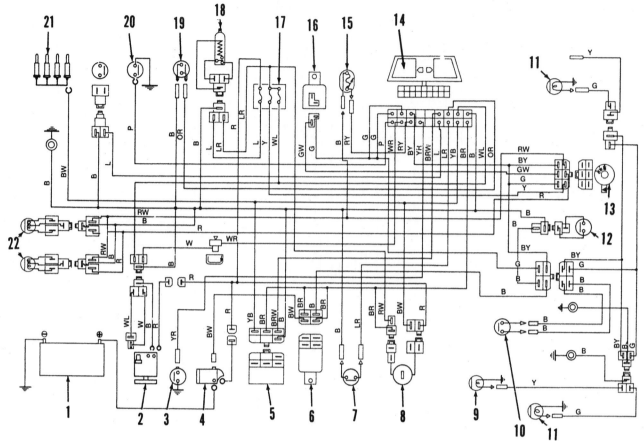

Fig. 266—Wiring diagram typical of Model 2120. Refer to Fig. 265 for legend.

B. Black
G. Green
L. Blue
P. Pink
R. Red
W. White
Y. Yellow
BR. Black/red
BRW. Brown/white
BW. Black/white
BY. Black/yellow
GW. Green/white
LR. Blue/red
OR. Orange/red
RW. Red/white
RY. Red/yellow
WL. White/blue
WR. White/red
YB. Yellow/black
YR. Yellow/red